Communications
in Computer and Information Science 499

More information about this series at http://www.springer.com/series/7899

Alexander Plakhov
Tatiana Tchemisova
Adelaide Freitas (Eds.)

Optimization
in the Natural Sciences

30th Euro Mini-Conference, EmC-ONS 2014
Aveiro, Portugal, February 5–9, 2014
Revised Selected Papers

 Springer

Editors
Alexander Plakhov
University of Aveiro
Aveiro
Portugal

Adelaide Freitas
University of Aveiro
Aveiro
Portugal

Tatiana Tchemisova
University of Aveiro
Aveiro
Portugal

ISSN 1865-0929 ISSN 1865-0937 (electronic)
Communications in Computer and Information Science
ISBN 978-3-319-20351-5 ISBN 978-3-319-20352-2 (eBook)
DOI 10.1007/978-3-319-20352-2

Library of Congress Control Number: 2015941859

Springer Cham Heidelberg New York Dordrecht London

Printed on acid-free paper

Springer International Publishing AG Switzerland is part of Springer Science+Business Media
(www.springer.com)

Preface

This Springer volume of *Communications in Computer and Information Science* is dedicated to the 30th EURO mini-conference on Optimization in the Natural Sciences (EmC-ONS 2014), which was held during February 5–9, 2014, in Aveiro, Portugal.

The conference attracted more than 100 registered participants who represented 21 countries from four continents. More than 70 contributed talks were divided into three streams—Optimization and Applications, Dynamical Systems, and Statistics, Bioinformatics, and Health Sciences—and constituted 22 sessions. The participants discussed recent achievements in optimization theory and related areas, exchanged experiences in solving real-world problems, and reported on the latest developments of appropriate models of optimization and their applications in the natural sciences. The 30th EURO mini-conference provided an excellent forum for researchers and practitioners in optimization to promote their recent advances to the wider scientific community and to identify new research challenges in theory, methods, and applications.

The conference topics reflected the huge diversity of different lines of research in optimization and its application in the natural sciences, including:

- Analysis of microarray data or next-generation sequencing
- Applications of modeling and optimization in physics, biology, chemistry, and medicine
- Billiard theory and applications
- Biomedical engineering
- Design optimization
- Data visualization for optimal decisions
- Image processing
- Infinite and semi-infinite optimization with applications
- Inverse problems
- Linear and nonlinear optimization and applications
- Multi-criteria optimization with applications
- Multi-scale optimization with applications
- Optimal control applied to biological models
- Optimal mass transfer
- Optimization in bioinformatics and computational biology
- Shape optimization
- Solution of optimization problems using statistical methods
- Statistics in high-dimensional data
- Statistical methods and visualization
- Statistical and probabilistic modeling
- Wave scattering

Based on a rigorous reviewing process realized by the members of the Program Committee, 13 papers were selected for publication in this volume. The keywords of the selected papers reflect the diversity of different lines of research in optimization and their applications in the natural sciences covered in this volume: optimal control, data visualization, spatial data analysis, shape optimization, billiards, multi-objective portfolio optimization, Markov chains, warehousing, multi-criteria optimization, simulation of information processing, principal component analysis in clustering, Herglotz's variational problems, multiple-response surface optimization, unreliable queueing systems, inverse problems, optimization of the hyperbolic type systems, suboptimal optimization, geometric optics, random access and others. The articles are grouped into three sections: Optimization and Applications, Dynamical Systems, and Modeling and Statistical Techniques for Data Analysis.

As guest editors, we would like to thank all the authors who contributed to this volume and all the reviewers who accepted the invitation to provide their expertise and give constructive comments. Our special thanks to the computer science editorial team at Springer, in particular to Aliaksandr Birukou, Frank Holzwarth, and Leonie Kunz for the opportunity to organize this volume, their expertise and coordination of the editorial process, and the continuous support and assistance.

March 2015

Alexander Plakhov
Tatiana Tchemisova
Adelaide Freitas

Organization

Program Chairs

Alexander Plakhov   University of Aveiro, Portugal
Tatiana Tchemisova   University of Aveiro, Portugal
Adelaide Freitas   University of Aveiro, Portugal

Program Committee

Adil Bagirov   University of Ballarat, Australia
Adilson Elias Xavier   Federal University of Rio de Janeiro, Brazil
Boris T. Polyak   Institute of Control Problems, Moscow, Russia
Domingos M. Cardoso   University of Aveiro, Portugal
Gerhard-Wilhelm Weber   Middle East Technical University, Turkey
Guiseppe Buttazzo   University of Pisa, Italy
János D. Pintér   Pintér Consulting Services, Inc., Canada
Joaquim Júdice   University of Coimbra, Portugal
Julius Žilinskas   Vilnius University, Lithuania
Leonidas Sakalauskas   Vilnius University, Lithuania
Lisete de Sousa   University of Lisbon, Portugal
Mourad Elloumi   University of Tunis-El Manar, Tunisia
Oliver Stein   Karlsruhe Institute of Technology, Germany
Purificación Galindo   University of Salamanca, Spain
Vladimir Bushenkov   University of Évora, Portugal
Miguel Pinheiro   University of St. Andrews, UK
Vadim Strijov   Moscow Institute of Physics and Technology, Russia

Contents

Dynamical Systems

Motion of a Rough Disc in Newtonian Aerodynamics

Sergey Kryzhevich[1,2](✉)

[1] Department of Mathematics, Center for Research and Development
in Mathematics and Applications, University of Aveiro, Aveiro 3810-193, Portugal
[2] Faculty of Mathematics and Mechanics, Saint-Petersburg State University,
28, Universitetskiy Pr., 198503 Peterhof, Saint-Petersburg, Russia
kryzhevicz@gmail.com, s.kryzhevich@spbu.ru

Abstract. Dynamics of a rough disc in a rarefied medium is considered.
We prove that any finite rectifiable curve can be approximated in the
Hausdorff metric by trajectories of centers of rough discs provided that
the parameters of the system are carefully chosen. To control the dynamics of the disc, we use the so-called inverse Magnus effect which causes
deviation of the trajectory of a spinning body. We study the so-called
response laws for scattering billiards e.g. relationship between the velocity of incidence of a particle and that of reflection. We construct a special
family of such laws that is weakly dense in the set of symmetric Borel
measures. Then we find a shape of cavities that provides selected law of
reflections. We write down differential equations that describe motions
of rough discs. We demonstrate how a given curve can be approximated
by considered trajectories.

Keywords: Billiards · Shape optimization · Magnus effect · Rarified
medium · Retroreflectors

1 Introduction

Consider a body with a piecewise smooth boundary moving in a two-dimensional
rarefied homogeneous medium. The particles composing this medium are initially
at rest. They never interact, they collide elastically with the body and move freely
between consecutive reflections from the boundary of the body.

This simple aerodynamic model was first introduced by Newton in his *Prin-
cipia* (1687). He studied a particular case of this model where a convex axially
symmetric body translates along its axis of symmetry. Due to collisions with particles of the medium, the force of resistance slows down the motion of the body.
Newton studied the problem of finding the shape of the body that minimizes
the force of resistance. The solution looks like a truncated cone with a slightly
inflated lateral surface. Several generalizations of Newton's problem related to
(generally) nonconvex and/or non-symmetric bodies have been studied in 1990s
and 2000s by various authors [1–10]. There are open problems in this area; for

© Springer International Publishing Switzerland 2015
A. Plakhov et al. (Eds.): EmC-ONS 2014, CCIS 499, pp. 3–19, 2015.
DOI: 10.1007/978-3-319-20352-2_1

instance, the shape of the convex and non-symmetric body of least resistance is not understood.

These investigations are closely related to the so-called problem of invisibility. One constructs a system of mirrors, invisible for an observer (observers), placed in a fixed point (points) or looking from a fixed direction (directions). Though the complete invisibility is impossible, some of related problems, for example, invisibility from one point or invisibility from one direction have been already solved [11,13].

Dynamics of a rarefied gas is a well-studied problem, see [14–19] and references therein.

Even more difficult and diverse are problems related to combined translational and rotational motion of bodies in a rarefied medium. Some of these problems are addressed in [20–27] under the assumption that the rotational motion is much slower than the translational one. In this case interaction of each individual particle with the body occurs as if there were no rotation at all: the turn of the body during the time of interaction can be neglected. It is shown, in particular, that the resistance of a convex body, in Euclidean space of arbitrary dimension, can be both increased and decreased by roughening its surface. The rates of maximum increase and decrease are found to depend only on the dimension and not on the original convex body; in the 3D case they are equal, respectively, to 2 and (approx.) 0.969445.

The Newtonian dynamics of a body that performs both translational and rotational motion is a very intriguing and completely unexplored subject even in the 2D case. Even attempts to study dynamics of very simple bodies, like a rod, not to say about an ellipse or a triangle, meet serious difficulties. The only exception is a circle, whose dynamics is trivial: the path of its center is a straight line. In this paper, we consider the so-called rough discs that represent the idea of a set, close to a ball in the Hausdorff metrics (see the beginning of Sect. 2). The principal goal of the article is the following.

We show that trajectories of centers of rough discs are dense in the set of finite rectifiable plane curves endowed with the Hausdorff metrics.

The proof of this statement is based on the following idea. In the typical case, if a disc rapidly rotates, say, counterclockwise, then the velocity vector of its center of mass changes in the clockwise direction. This phenomenon is called the *inverse Magnus effect*, see Fig. 1. The word "inverse" means this effect is inverse to the Magnus effect proper for classical gas dynamics and well-known for soccer or ping-pong players where a ball deviates at the direction of rotation. There is no contradiction: influence of a classic gas is very different from one of rarefied media.

The magnitude of the effect depends on the shape of cavities on the boundary of the disc and on the relative angular velocity λ of the disc. In this paper we construct a very special cavity in such a way that (i) the relative angular velocity monotonously increases and (ii) the magnitude of the effect is nearly zero for all values of λ except for several (relatively small) intervals of values. On these intervals the effect is adjusted so as to ensure right turn of the velocity vector to a certain angle.

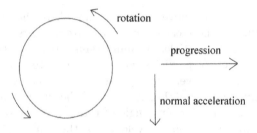

Fig. 1. Inverse Magnus effect.

Therefore, the basic idea of the proof of our main results is to use shapes of cavities to control inverse Magnus effect. We believe that our construction can be generalized to three dimensions, but postpone the 3D study to the future.

Structure of the Paper. In Sect. 2, we consider an immobile scattering billiard which gives a simplified model for dynamics of a particle inside a hollow. In the next section we formulate physical assumptions on a moving body and medium and introduce some notions. We introduce the concept of δ – pseudotrajectory, corresponding to immobile billiard system. We show that, for sufficiently precise approximations to so-called perfect rough discs, scattering billiard model gives a good approximation for relative motions of particles inside cavities. Then, in Sect. 4, we can apply the model for motions of rough bodies [12]. We study some special types of cavities and related reflection laws (Sects. 5–7). We formulate the main result of the paper (Sect. 8) and prove it. The main idea of the proof is approximation of a curve by broken lines, for which we can write down equations of motions and shapes of cavities explicitly.

2   Laws of Scattering for Immobile Billiards

We start with the definition of a rough disc. Fix $r > 0$, take a regular n_0-gon ($n_0 \geq 3$) inscribed in a circle with radius r (let its center be O), and replace each side of the n_0-gon with a curve joining its endpoints. Each curve is piecewise smooth, does not have any self-intersections, and is contained in the circular sector with vertices at O and at the endpoints of the corresponding side. In addition, all the curves are congruent: each curve can be obtained from another one by rotation around O by $2\pi k/n_0$. For each integer $n > n_0$ make a similar procedure: take a regular n-gon inscribed in the same circle and replace its sides with curves, so that the obtained sequence of sets tends to the circle in Hausdorff metrics. The union of the curves in each n-gon bounds a domain B_n.

Definition 1. The sequence of domains B_n, $n \geq n_0$ is called a *rough disc*.

Thus, a rough disc is an idealized object. It is not a domain, but rather it can be informally viewed as the "limit" of a sequence of domains B_n. Its "boundary" is obtained by repetition of identical infinitesimal curves similar to the original one. They are interpreted as infinitesimal hollows on the disc boundary. The billiard scattering by the rough disc is uniquely defined by the shape of the curve.

We assume that the dics moves in a medium where the mass is uniformly distributed according to the measure m: $dm = \rho \, dS$, where S is the Lebesgue measure in $\mathbb{R}^2$. We treat particles as infinitesimal parts of the medium. We neglect Brownian motion of particles when we calculate interactions between particles and the body. However, we suppose that particles instantly fill in the space after the body was passing. Let $\boldsymbol{X} = (X, Y)$ be the current position of the center of the body, ϕ be the current angle of rotation of the body with respect to its initial position. Let $\boldsymbol{V}$ be the velocity of the center, $|\boldsymbol{V}| = V$. Denote by ω the angular velocity, by r be the radius of the disc and by $I = \kappa M r^2$ the moment of inertia. For a regular disc with uniformly distributed mass we have $I = M r^2 / 2$, and in any case $\kappa \in [0, 1]$. We introduce the angular coordinate ξ on the boundary of the ordinary disc representing the smooth approximation of the moving body, identifying this boundary with the unit circle $S^1 = [-\pi, \pi]/\{\pi = -\pi\}$. Recall the notion for the dimensionless relative angular velocity $\lambda = \omega r / V$. The force of resistance of the medium acting on the disc and the moment of this force are defined as limits, when $n \to \infty$, of the force and the moment of force acting on B_n. Using these values, we derive the equations of motion of a rough disc on the plane. These equations, and therefore the trajectory of the disc, depend on the shape of the infinitesimal curve forming its boundary. The natural question arises: which curves can be traversed by the disc center?

Description of scattering by a rough disc and equations of motions for such discs can be found in [27], and in Chaps. 4 and 7 of the book [12]. We partly reproduce them here.

Definition 2. A *hollow* is a piecewise smooth non self-intersecting curve contained in a closed isosceles triangle whose base is the segment joining the endpoints of the curve. The segment is called the *opening* of the hollow.

We use the notion Ω for a hollow and I for its opening. Introduce the uniform coordinate $\xi \in [0, 1]$ on the opening I; the values $\xi = 0$ and $\xi = 1$ correspond to its endpoints. Let $\boldsymbol{n}$ be the unit outer normal to I. Consider a particle that enters a hollow Ω through its opening I. Fix the point ξ where it intersects the opening and the angle $\varphi \in (-\pi/2, \pi/2)$ formed by $-\boldsymbol{n}$ and the incidence velocity $\boldsymbol{v}$. If the particle makes a finite number of reflections from regular points of Ω, intersects I again and leaves, we denote by $\xi^+ = \xi_\Omega^+(\varphi, \xi)$ the point of the second intersection and by $\varphi^+ = \varphi_\Omega^+(\varphi, \xi)$ the angle formed by $\boldsymbol{n}$ and the velocity $\boldsymbol{v}^+$.

Almost all particles leave the hollow Ω after a finite number of reflections. This follows from the measure-preserving property of billiard and from Poincaré's recurrence theorem. Thus for almost all initial conditions $(\varphi, \xi) \in [-\pi/2, \pi/2] \times [0, 1]$ the values $\varphi_\Omega^+(\varphi, \xi)$ and $\xi_\Omega^+(\varphi, \xi)$ are well-defined. Introduce the probability measure μ on $[-\pi/2, \pi/2] \times [0, 1]$ according to $d\mu(\varphi, \xi) = \frac{1}{2} \cos \varphi \, d\varphi \, d\xi$. The map $T_\Omega : (\varphi, \xi) \mapsto (\varphi_\Omega^+(\varphi, \xi), \xi_\Omega^+(\varphi, \xi))$ is defined on a full-measure subset of $[-\pi/2, \pi/2] \times [0, 1]$ and maps it bijectively onto itself. Moreover, it preserves the measure μ and is involutive, $T_\Omega = T_\Omega^{-1}$.

Next introduce the Borel measure η_Ω on the square $\square := [-\pi/2, \pi/2] \times [-\pi/2, \pi/2]$ as follows: $\eta_\Omega(A) = \mu(\{(\varphi, \xi) : (\varphi, \varphi_\Omega^+(\varphi, \xi)) \in A\})$ for any Borel set $A \subset \square$.

This measure can be defined in a different way: let σ_Ω be the mapping $(\varphi, \xi) \mapsto (\varphi, \varphi_\Omega^+(\varphi, \xi))$ from $[-\pi/2, \pi/2] \times [0, 1]$ to $\square$; then η_Ω is the measure $\eta_\Omega = \sigma_\Omega^\# \mu$. Here $\sigma_\Omega^\#$ is the mapping of measures induced by σ_Ω.

Definition 3. η_Ω is called the *measure induced by the hollow* Ω.

Define the probability measure γ on $[-\pi/2, \pi/2]$ by $d\gamma(\varphi) = \frac{1}{2} \cos\varphi\, d\varphi$. For a set $A \subset \square$, denote $A^* = \{(\varphi, \varphi^+) : (\varphi^+, \varphi) \in A\}$. Denote by Υ the set of Borel measures η on $\square$ such that for all Borel sets $A \subset \square$ and $I \subset [-\pi/2, \pi/2]$ one has $\eta(A) = \eta(A^*)$ and $\eta(I \times [-\pi/2, \pi/2]) = \gamma(I)$. The fact that $\eta_\Omega \in \Upsilon$ can be easily deduced from the measure preserving and involutive properties of the map T_Ω; see [12] for details. The following important theorem states that, inversely, the set of measures induced by hollows is weakly dense in Υ.

Density Theorem [12]. *The set* $\{\eta_\Omega : \Omega$ *is a hollow*$\}$ *is weakly dense in* Υ. *In other words, for any* $\eta \in \Upsilon$ *there exists a sequence of hollows* Ω_k *such that*

$$\lim_{k\to\infty} \iint_\square f(\varphi, \varphi^+)\, d\eta_{\Omega_k}(\varphi, \varphi^+) = \iint_\square f(\varphi, \varphi^+)\, d\eta(\varphi, \varphi^+)$$

for any continuous function $f : \square \to \mathbb{R}$.

3   Pseudotrajectories

Given a $\delta > 0$, we introduce the concept of a δ-*pseudotrajectory* for a billiard.

Definition 4. We say that a piecewise C^1 smooth curve $\boldsymbol{x}(t) : t \in [t_0, \hat{t}_0]$ is a δ-pseudotrajectory for the exterior billiard corresponding to an immobile body A if the following statements are true.

1. $\boldsymbol{x}(t) \notin \operatorname{int} A$ for all $t \in [t_0, \hat{t}_0]$.
2. The set of t such that $\boldsymbol{x}(t) \in \partial A$ is finite. Let it be $\{t_1, \ldots, t_N\} : N \in \mathbb{N} \bigcup \{0\}$. We also use the notation $t_{N+1} = \hat{t}_0$.
3. For all $k \in \{1, \ldots, N\}$ the velocities $\boldsymbol{v}_{r+} = \boldsymbol{v}(t_k + 0) = \dot{\boldsymbol{x}}(t_k + 0)$ and $\boldsymbol{v}_{r-} = \boldsymbol{v}(t_k - 0) = \dot{\boldsymbol{x}}(t_k - 0)$ of the corresponding impacts satisfy inequalities

$$|\boldsymbol{v}_{e+} - \boldsymbol{v}_{e-} - 2\langle \boldsymbol{v}_{e-}, \boldsymbol{n}\rangle \boldsymbol{n}| \leq \delta. \tag{1}$$

   If $x(t)$ is a singularity point, we select one of two possible values for normal vectors.
4. The function $\boldsymbol{v}(t) = \dot{\boldsymbol{x}}(t)$ is piecewise smooth and $|\boldsymbol{v}(t) - \boldsymbol{v}(t_k + 0)| \leq \delta$ for any $k \in \{0, \ldots, N\}$ and any $t \in (t_k, t_{k+1})$.

We use this notion to describe trajectories of particles of non-zero mass that interact with a moving and rotating body.

Definition 5. A rough disc defined by a sequence B_n is *perfect* if there exist $m_0 \in \mathbb{N}$, $\lambda_0 > 0$, $0 < \delta < \pi/2$ and $K > 0$ such that for any $m \geq m_0$ all δ-pseudotrajectories, entering the corresponding hollow with the incident angle $\geq \lambda_0$, have at most K impacts before they leave the hollow.

We make the following assumptions on interactions between the body and particles.

1. If a particle collides once with a point of the boundary of the body out of any hollow, we neglect all later interactions between the particle and the body.
2. We assume that there is a number $K > 0$ such that all particles interacting with a fixed hollow of the body, have at most K impacts and leave the hollow.

Direct calculations lead us to the following statement.

Lemma 1. *Let $\{B_n\}$ be a rough disc, h_n be diameters of corresponding cavities. Let a_0, v_0, v_1 and ω_0 be positive constants. Suppose that a body B_n translates and rotates during a period $[0, T]$ so that $|\dot{X}(t)| \in [v_0, v_1]$, $|\ddot{X}(t)| \le a_0$. Here $X(t)$ is the position of the center of the body, ω is the angular velocity. Then there exist $n_0 \in \mathbb{N}$ and $C_1, C_2 > 0$ such that any particle entering a cavity with incidence angle $\le \lambda_0$ spends at most $C_1 h_n$ units of time inside the cavity. The part of the trajectory inside a cavity forms a $C_2 h_n$ pseudotrajectory.*

4 Dynamics of Perfect Rough Bodies

The dynamics of the rough disc is described by the following system of ordinary differential Eq. [12, Theorem 7.1, p. 203]:

$$M \dot{V} = R(\eta, \omega, V) = \frac{8}{3} r\rho V^2 \overline{R}(\eta, \lambda); \qquad I\dot{\omega} = R_I(\eta, \omega, V) = \frac{8}{3} r\rho V^2 \overline{R_I}(\eta, \lambda). \quad (2)$$

Here η is the billiard law corresponding to the selected rough disc. Formulae for dimensionless resistances $\overline{R}$ depend on λ. Here we assume that $\lambda > 1$. Consider the coordinate system associated with the vector V and the orthogonal vector $V^\perp$. Functions $\overline{R}$ and R_I can be found from the following formulae:

$$\overline{R}(\eta, \lambda) = (R_T(\eta, \lambda), R_L(\eta, \lambda)); \quad R_T(\eta, \lambda) = \int_\square c_T(x, y, \lambda) \, d\eta(x, y);$$

$$R_L(\eta, \lambda) = \int_\square c_L(x, y, \lambda) \, d\eta(x, y); \quad R_I(\eta, \lambda) = \int_\square c_I(x, y, \lambda) \, d\eta(x, y). \quad (3)$$

Here

$$c_T(x, y, \lambda) = \frac{3\cos\frac{x-y}{2}}{\sin\zeta}((\lambda^3 \sin^3 x + 3\lambda \sin x \sin^2 \zeta)\cos\zeta\cos\frac{x-y}{2} -$$
$$(3\lambda^2 \sin^2 x \sin\zeta + \sin^3 \zeta)\sin\zeta\sin\frac{x-y}{2})\chi_{x \ge x_0}(x, y);$$

$$c_L(x, y, \lambda) = -\frac{3\cos\frac{x-y}{2}}{\sin\zeta}((\lambda^3 \sin^3 x + 3\lambda \sin x \sin^2 \zeta)\cos\zeta\sin\frac{x-y}{2} + \quad (4)$$
$$(3\lambda^2 \sin^2 x \sin\zeta + \sin^3 \zeta)\sin\zeta\cos\frac{x-y}{2})\chi_{x \ge x_0}(x, y);$$

$$c_I(x, y, \lambda) = -\frac{3}{2}\frac{\lambda^3 \sin^3 x + 3\lambda \sin x \sin^2 \zeta}{\sin\zeta}(\sin x + \sin y)\chi_{x \ge x_0}(x, y);$$

$\zeta = \arcsin\sqrt{1 - \lambda^2 \cos^2 x}$, $x_0 = \arccos(1/\lambda)$; χ stands for the characteristic function. Applying the mentionned Theorem 7.1, we use the result of Lemma 1.

Make a transformation of variables in Eq. (2). First of all, select τ so that:

$$d\tau = \frac{8r\rho V}{3M} \, dt. \quad (5)$$

Then we define θ so that $\boldsymbol{V} = V(\cos\theta, \sin\theta)$. Let $\beta = Mr^2/I = \kappa^{-1}$ be the inverse relative moment of inertia of the rough disc. It follows from Eq. (2) that

$$\frac{d\lambda}{d\tau} = \beta R_I(\lambda) - \lambda R_L(\lambda), \quad \frac{dV}{d\tau} = -V R_L(\lambda), \quad \frac{d\theta}{d\tau} = -R_T(\lambda). \tag{6}$$

Observe that the variable τ is a natural parametrization of the trajectory of the center of the disc. Namely, if $S(t)$ is the path passed by the center of the disc by the moment t then $dS/d\tau = 3M/(8\rho r) = \text{const}$.

In following three sections we provide a family of specially selected roughnesses and justify that the proposed model of dynamics is applicable for rough discs with such shapes of cavities. First of all we need two types of auxiliary scattering billiards.

5 Bunimovich Mushroom

Let us introduce the so-called retroreflectors. Consider a family of domains $\Theta_h \subset \mathbb{R}^2$ (h is a small positive parameter) with a piecewise smooth boundary $\partial\Theta_h$ which can be represented as a disjoint union $\partial\Theta_h = \Omega_h \bigcup I_h$ where Ω_h and I_h satisfy following properties.

1. The arc Ω_h is a hollow with the opening I_h.
2. Consider a uniform distribution of pairs $(x_-, \nu_-) \in I_h \times [-\pi/2, \pi/2]$ that is the point and the angle of incidence. Let ν_+ be the angle of the last intersection between the trajectory of a particle and the segment I_h, then for any $\sigma > 0$ the proportion of particles such that $|\nu_+ + \nu_- - \pi| > \sigma$ tends to zero as $h \to 0$.

In this paper we consider so-called "Bunimovich mushroom" [28,29], Fig. 2. There exist other patterns of retroreflectors [12], Chap. 9].

The pattern of the mushroom, we use in this article is the following: a domain Θ_h which is a union two domains: Θ_{h1} and Θ_{h2}. The first one ("pileus" of the mushroom) is a strictly convex domain which is the upper part of an ellipse, whose principal axis is horizontal. The second part of the mushroom (call it stipe) is a $b_{12} \times b_{13}$ rectangle. Let b_{11} be the length of the long axis of the ellipse Θ_{h1}. We assume that the tops of the stipe coincide with the foci of the ellipse. Suppose that

$$b_{12}/b_{11} = 2h; \qquad b_{13}/b_{12} = h. \tag{7}$$

We call this value h imperfectness of the mushroom.

We claim that $I_h = (P_L P_R)$ on Fig. 2 is the opening for the considered scattering billiard. Consequently, we suppose that Ω_h is the rest of the boundary of the "mushroom".

Let us prove that this mushroom is a retroreflector. Fixed $\sigma > 0$ we define sets $\Sigma_\sigma = \{(\boldsymbol{x}_-, \boldsymbol{v}_-) \in X : |\boldsymbol{x}_+ - \boldsymbol{x}_-| \leq \sigma b_{22}, \; |\boldsymbol{v}_+ + \boldsymbol{v}_-| \leq \sigma\}$.

Lemma [12, Lemma 4.1, p. 115]. *For any $\sigma > 0$ there exists a $h_0 > 0$ such that if $h \in (0, h_0)$ and conditions (7) are satisfied, the measure of the set Σ_σ is greater than $1 - \sigma$.*

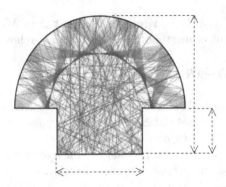

Fig. 2. Mushroom billiard.

6   Amphora Billiard: A Quasi-Elastic Hollow

Select a small positive parameter h called imperfectness of the billiard. Consider two arcs of confocal parabolas given by equations $x = \pm(1-y^2)/2$, $y \in [0,1]$. Link the lower ends of these arcs by a segment. We obtain a curvilinear triangle. Cut the middle part $F_L F_R$ of the base of this triangle corresponding to $x \in [-h,h]$. Construct two segments $A_L F_L$ and $A_R F_R$ of the length h^2 at ends of the obtained gap, which make angles $\pm\pi/4$ with the axis Ox, Fig. 3(a)). The amphora domain is constructed.

Later on we deal with modifications of amphora billiards. We take the parameter b_{21} so that

$$b_{21} = h; \qquad b_{22}/b_{21} = 2h; \qquad b_{23}/b_{22} = h \qquad (8)$$

where h is the imperfectness, b_{22} is the width of the entrance corridor of the billiard domain, call it "neck", b_{23} is the length if this corridor.

Let $X = [-b_{22}, b_{22}] \times (-\pi, 0)$ be endowed with the smooth measure ν with the density $d\nu = -\sin v_-/(4b_{22})dx\, dv_-$.

Next lemma demonstrates that this amphora hollow works like a smooth mirror i.e. for "almost" all particles the angle of incidence "almost" equals to the angle of reflection. Let N_σ be the set of initial conditions $(\boldsymbol{x}_-, \boldsymbol{v}_-) \in X$ of the entrance which correspond to billiard trajectories with two impacts such that $|\boldsymbol{x}_+ + \boldsymbol{x}_-| \leq \sigma b_{22}$, $|\boldsymbol{v}_+ - R\boldsymbol{v}_-| \leq \sigma$. Here $R(v_x, v_y) = (v_x, -v_y)$ and, as usually, unit vectors $\boldsymbol{v}_\pm$ correspond to angles $v_\pm$.

Lemma 2. *For any $\sigma > 0$ there exists a $h_0 > 0$ such that if $h \in (0, h_0)$ and conditions (8) are satisfied, the Lebesgue measure of the set N_σ is greater than $1-\sigma$.*

Proof. Let (x_-, v_-) be the initial position and the angle of the initial velocity of a particle. We identify v_- with a point of the lower semicircle. Let (x_+, v_+) correspond to the exit of the particle. Here v_+ is a point of the upper semicircle.

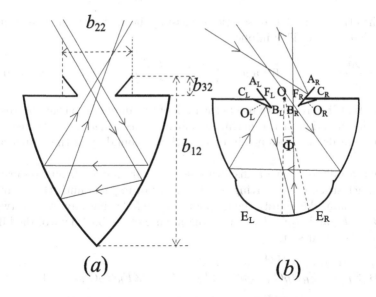

Fig. 3. Amphora billiard (a) and its modification (b).

Observe that any particle, corresponding to initial conditions $(0, v_-)$, $|\tan v_-| > 2h$, is reflected back to the same point after two impacts (unless the particle is moving strictly down). Moreover, after the first impact the motion of the particle is strictly parallel to the axis Ox. Let v_0 and v_1 be such that $\tan v_0 = -2h$, $\tan v_1 = 2h$. Then there exists a $\delta > 0$ such that every trajectory of the amphora billiard, corresponding to initial conditions (x, v): $|x| < \delta$, $v \in (v_-, v_+)$, $v \neq -\pi/2$ has exactly two impacts and both of them correspond to points of "sides of the amphora" i.e. parabolas. It suffices to prove that

$$D_v = \frac{\partial(x_+, v_+)}{\partial(x_-, v_-)}(0, v) = \begin{pmatrix} d_{11} & d_{12} \\ d_{21} & d_{22} \end{pmatrix} = \begin{pmatrix} \pm 1 & 0 \\ 0 & -1 \end{pmatrix}$$

for any $v \in (\Theta_-, \Theta_+)$. The sign of the element d_{11} is not important for us.

Since every trajectory that passes via the focus comes back to the focus after two reflections, we have $d_{12} = 0$. Due to symmetry reasons, $d_{22} = -1$, $d_{11} = \pm 1$.

Let n_- and n_+ be unit normal vectors for points of the first and the second impact respectively. Let (x_-, v_-) be initial conditions for the trajectory. Then $n_\pm$ are functions of x_- and, moreover, grace to the structure of the considered domain, the vector n_- uniquely defines the point of the first impact and, consequently, uniquely defines the vector n_+. Let $n_\pm$ be the angles between $n_\pm$ and Ox. Consider the angle α between the axis Ox and the trajectory of the particle after the first impact. Clearly, $\alpha = 0$ for all solutions, passing via the focus. Due to reflection law, $\alpha = v_- - 2n_-$. Comparing the trajectory of a particle with one obtained by reversion of time we get $v_- - 2n_- + v_+ - 2n_+ = \pi$.

On the other hand, for all solutions, passing via the focus, one can easily see that $dn_+/dn_- = -1$. This implies

$$\left(\frac{\partial n_+}{\partial x_-} + \frac{\partial n_-}{\partial x_-}\right)\bigg|_{x_-=0} = 0 \quad \text{and} \quad \left(\frac{\partial v_+}{\partial x_-} + \frac{\partial v_-}{\partial x_-}\right)\bigg|_{x_-=0} = 0 \quad \Rightarrow d_{21} = 0. \quad \square$$

Note also, that if a trajectory meets the neck of the amphora so that the absolute value of the direction of the entrance velocity is less than $\pi/4$, the particle is reflected upwards and does not interact with the boundary of the amphora any more.

Amphora billiards have a disadvantage, similar to one of mushrooms: particles can get stuck there, having a big number of impacts until they leave the amphora domain. We modify the amphora in the following way. Attach two triangles $B_L C_L F_L$ and $B_R C_R F_R$ to horizontal parts of the boundary of the billiard (Fig. 3(b)). We do it so that

1. $|B_L F_L| = |B_R F_R| = h^{5/4} \quad (= o(F_L F_R))$,
2. $\angle F_L B_L C_L = \angle F_R B_R C_R = \pi/6$, $\angle F_L B_L C_L = \angle F_R B_R C_R = \pi/4$.

We introduce a coordinate $\Phi \in [-\pi/2, \pi/2]$ on boards of the amphora. This coordinate corresponds to the inclination of the line, passing through the origin and the selected point. Consider two symmetric points O_L and O_R that are centers of segments $[B_L F_L]$ and $[B_R F_R]$ respectively. Replace parts of parabolas, corresponding to $\Phi \in [-\pi/4, \pi/4]$ with arcs of ellipses E_L and E_R such that one focus for both of these ellipses is O and another one is O_L for E_R and O_R for E_L. The modified amphora domain is constructed, Fig. 3(b).

Now we study billiard trajectories for the modified amphora billiard. Suppose that the angle between the initial velocity and the line is less than $\pi/7$.

If a particle hits the boundary at one of points of $[A_L B_L]$ or $[A_R B_R]$ it is reflected upwards and does not have any other impacts. Otherwise, it interacts twice with arcs of parabolas. After that, due to Lemma 2 there exist following three alternatives, Fig. 3.

1. A particle leaves the amphora domain forever without having any more impacts.
2. A particle hits $[A_L B_L]$ or $[A_R B_R]$ and leaves the amphora domain.
3. A particle hits $[B_L C_L]$ or $[B_R C_R]$ then maps to a point of E_L or E_R respectively. After that, the trajectory crosses the $2h^{5/4}$ – vicinity of the origin and leaves the amphora domain forever.

Lemma 2 guarantees that the "majority" of trajectories behave according to the first scenario. Note that for any initial conditions of the considered type the number of impacts cannot exceed 4.

7 Hybrid Hollows

Now we are ready to construct the rough element, i.e. the hollow, corresponding to the rough disc with a prescribed law of reflection. We modify the amphora

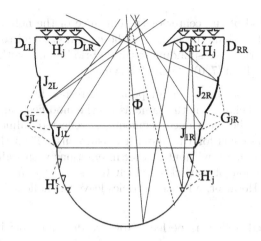

Fig. 4. Hybrid billiard.

billiard so that for some selected directions of incident particles it works as a retroreflector and for some others it works as a quasielastic reflector.

Select two symmetric sets of non-intersecting segments J_{kL} and J_{kR} ($k = 1, \ldots m$) given by $J_{kR} = [\Phi_k^0, \Phi_k^1]$, $J_{kL} = [-\Phi_k^1, -\Phi_k^0]$. Assume that $5\pi/14 = \pi/2 - \pi/7 < \Phi_1^0 < \Phi_1^1 < \ldots < \Phi_m^0 < \Phi_m^1 < \pi/2$.

Given a point D_{RR} (the right edge of the hollow) we attach an arc of the ellipse with the foci at F_L and F_R and corresponding to $\Phi \in (\Phi_m^0, \Phi_m^1)$. Then we draw an arc of the parabola with a focus at the origin and the vertical axis of symmetry through the free end of the constructed arc of the ellipse. We do it for $\Phi \in (\Phi_{m-1}^1, \Phi_m^0)$. We repeat similar constructions of arcs of ellipses with same foci and parabolas with the same focus until we reach $\Phi = \Phi_1^1$. Then we attach the last arc of parabola, corresponding to $\Phi \in (\pi/4, \Phi_1^1)$. To finish the construction we attach an arc of an ellipse, corresponding to $(\pi/4, \pi/2)$ similarly to what we did for modified amphora billiards (Figs. 3 and 4).

It may happen that a trajectory or a pseudotrajectory which hits a parabolic part of the boundary near its junction with an elliptic part, next hits an elliptic part on the opposite side of the hollow. Generally, this means that the corresponding billiard trajectory hits one of segments $\Sigma_L = [D_{LL}, D_{LR}]$ or $\Sigma_R = [D_{RL}, D_{RR}]$ on the upper part of the boundary of the hollow (Fig. 4). Let $G_{1L}, \ldots, G_{mL}$ and $G_{1R}, \ldots, G_{mR}$ be junctions between elliptic and parabolic sectors. Consider $H_1, \ldots H_{2m}$ that are points on the union $\Sigma_L \bigcup \Sigma_R$, corresponding to "parabolic+elliptic" reflections from points G_{kL} and G_{kR} or vice versa. We put a system of flat mirrors (segments) of sizes $h^{5/4}$ centered at H_j ($j = 1, \ldots, 2m$) so that all $h^{3/2}$ pseudotrajectories, hitting first parabolic, then elliptic sectors, are reflected via these mirrors to $h^{9/8}$ neighborhoods of points $H_1', \ldots H_{2m}'$ such that $H_j' \in (-\pi/4, -\pi/7) \bigcup (\pi/7, \pi/4)$ for all j. We put flat mirrors of lengths $h^{9/8}$, centered at points H_j' so that all considered trajectories and pseudotrajectories are reflected by these mirrors to the

$h^{17/16}$ neighborhood of the center of the entrance of the hollow (Fig. 4). That size is still much less than the length of the entrance, equal to h. Trajectories and pseudotrajectories corresponding to this hybrid billiard, with incident angles $v_- \in (-\pi, -6\pi/7) \bigcup (-\pi/7, 0)$ are the following.

1. If a pseudotrajectory does not hit points, corresponding to one of segments J_{kL} or J_{kR}, the behavior is the same as for the modified amphora billiard.
2. If it hits one of the mentioned segments, is reflected "almost back" (similarly to what happens for Bunimovich mushrooms). Then the *pseudotrajectory* leaves the domain without farther interactions with walls.
3. A small proportion of particles (which tends to 0 as $h \to 0$) has a distinct behavior. However, all such particles leave the hollow, having at most 4 impacts.

So, the constructed hollow is perfect. Now we describe how it is possible to cover almost all segment $I \in Ox$ (we may also do the same if I is an arc of the circle) with tops of hybrid billiards. Cut the middle part of I of the length $2h|I|$ and insert there a hybrid billiard of imperfectness h and the basis of the neck equal to $2h|I|$. Call this hollow one of the first generation. Let b_1 be the corresponding rescaling coefficient. Take $b_2 = \varrho h^2 b_1$. Here $\varrho < 1$ is the principle rescaling for smaller mushrooms of the "second generation", Fig. 4). Then we put $N \sim h^{-1}|I|$ non-intersecting hollows of the second generation whose tops correspond to subsegments of I. We repeat this procedure, creating hollows of the third level and so on. On the step number L, the measure of the part of the segment I, not covered by tops of already constructed hollows can be estimated by the value $|I|(1 - \tilde{h}/2)^L$. In the limit, we get a zero-measure Cantor set. However, we stop after finitely many steps.

8   Main Result

Theorem 1. *Let* $g : [a, b] \to \mathbb{R}^2$ *be a continuous rectifiable curve. Then for any* $\varsigma > 0$ *there exists a motion* $(\boldsymbol{X}(\tau), \varphi(\tau))$ *of a rough disc of radius* $r > 0$ *such that after a continuous and monotone increasing change of parameter* $\tau = \tau(t)$, $t \in [a, b]$ *one has*

$$|\boldsymbol{g}(t) - \boldsymbol{X}(\tau(t))| < \varsigma. \tag{9}$$

Here $\boldsymbol{X}(\tau)$ *is the position of the center of the disc;* $\varphi(\tau)$ *is the turn of the disc.*

Note that we the curve g is not necessarily injective: self-intersections and even coincidence of some fragments of the curve are allowed.

The following auxiliary theorem states that any broken line can be approximated by trajectories of rough discs. Namely, let $\boldsymbol{G}(t)$, $t \in [a, b]$ be a parameterized broken line with a finite number of segments, $\Gamma = \{\boldsymbol{G}(t) : t \in [a, b]\}$. Self-intersections are allowed, but we require that no vertex of the broken line is a point of intersection. Moreover, we approximate broken lines so that inclinations of every segment with respect to the previous one varies from $-\pi/4$ to 0. For instance, instead a rotation by the angle $\pi/4$, we apply seven rotations by $-\pi/4$.

Theorem 2. *For any $\varsigma > 0$ there exists a motion of a rough disc of radius $r > 0$ whose center is $\boldsymbol{X}(\tau)$ such that after a continuous and monotone increasing transformation $\tau = \tau(t)$, $t \in [a, b]$ inequality (9) is satisfied.*

Theorem 1 is an obvious consequence of Theorem 2. Indeed, each rectifiable curve can be uniformly approximated by broken lines, and each broken line can be uniformly approximated by trajectories of rough discs.

Proof of Theorem 2. First we notice that a curve homothetic to a trajectory of a rough disc is also a trajectory of a rough disc. Let $\boldsymbol{X}(t)$ be the motion of the center of a rough disc of radius r. Let $\omega(t)$ be its angular velocity and ϵ be a positive constant. Then the coordinate of the center of a disc of radius νr homothetic to the original one moving in the same medium with the initial velocity $\epsilon \boldsymbol{X}'(0)$, the initial angular velocity $\omega(0)$, is given by $\epsilon \boldsymbol{X}(t)$, and its angular velocity is $\omega(t)$.

This scaling argument allows one to reduce Theorem 2 to the problem of approximation of a broken line $\frac{1}{\epsilon}\boldsymbol{g}(t)$ where ϵ is a small parameter. Select a splitting of the broken line into segments with ends, corresponding to $a = T_0 < T_1 < \ldots < T_{m-1} < T_m = b$.

Take a disc B_{n_ε} with the roughness of the considered form. Introduce the measure in $[-\pi/2, \pi/2] \times [-\pi/2, \pi/2]$ associated with the cavity which has the density

$$\frac{1}{2}\cos x\{\delta(x - y) \cdot \chi_{J \cup J'}(x) + \delta(x + y) \cdot [1 - \chi_{J \cup J'}(x)]\}\, dx\, dy. \qquad (10)$$

if $|x|, |y| \leq 5\pi/14$. Here

$$J = \bigcup_{i=1}^{m} J_{iR} = \bigcup_{i=1}^{m}[\pi/2 - e^{-T_i/\varepsilon}, \pi/2 - e^{-(T_i + \Delta T_i)/\varepsilon}] \quad \text{and} \quad J' = -J, \qquad (11)$$

$i = 1, 2, \ldots, m$ is a finite set of indices. We select J_{iR} as "elliptic" segments on the boundary of a cavity (see Sect. 7 and Fig. 4). The initial angular velocity $\lambda(0) = \omega(0)/rV(0)$ is taken to be $\lambda(0) = e^{T_0/\varepsilon}$. $T_i - T_{i-1}$ is the length of the i – th segment of the broken line, $\Delta T_i = \varphi_i e^{-T_i/\varepsilon}$. Here φ_i are parameters, close to angles $\varphi_i^0 \in [-\pi/4, 0]$ between the i-th and $(i+1)$-th segments of the broken line. Now we note that $\varepsilon > 0$ is taken so small that all segments J_i are disjoint.

As the disc moves, the relative angular velocity increases and less of the part of the cavity is "observable" by particles. Depending on the value of λ either $J \bigcup J'$ or completion of this set dominate in the "observable" part. Respectively, we have rotation or "almost straight forward" motion. A small part ε of the boundary is filled with cavities. The rest, $1 - \varepsilon$, of the boundary is not filled, that is, is just a union of arcs of the unit circumference. Both parts are uniformly distributed along the boundary.

Consider the natural parametrization $\boldsymbol{g}(\tau)$, $\tau \in [T_0, T_m]$, where $[T_{j-1}, T_j]$ parameterize segments of the broken line. We find a motion of a rough disc of unit radius where $\boldsymbol{X}(\tau)$ is the position of the center and values S_j $(j = 0, \ldots, m)$, $\tau \in [S_0, S_m]$ so that

$$|\boldsymbol{g}(\tau)/\varepsilon - \boldsymbol{X}(\tau/\varepsilon)| < (m + 1)/\sqrt{\varepsilon} \qquad (12)$$

or, equivalently, $|g(\tau) - \varepsilon X(\tau/\varepsilon)| < (m+1)\sqrt{\varepsilon}$. Given ς, we take ε so that $\varsigma = (m+1)\sqrt{\varepsilon}$. Then we take the rescaling parameter $\kappa = \varepsilon$ and easily obtain inequality (9).

The motion of the disc is described in terms of the parameter τ proportional to the natural one (see (5)). It can be deduced from Eqs. (6) and (12) and from equations defining the measure (10), (11) that the differential equation for $\lambda(\tau)$ takes the form $\lambda' = \lambda u(\lambda, \varepsilon, \tau)$ where $u(\lambda, \varepsilon, \tau) \rightrightarrows 1$ as $\varepsilon \to 0$. So $\lambda = e^{w(\varepsilon, \tau)}$ where w is increasing with respect to τ and $w(\varepsilon, \tau)/\tau \rightrightarrows 1$. Consider values S_j defined by equalities $w(\varepsilon, S_j) = T_j/\varepsilon$.

Using Eqs. (3), (4) and (6), introduce the notation $x_0 = x_0(\lambda) = \arccos(1/\lambda)$, and obtain the equality

$$\frac{1}{2} \int_{x_0}^{\pi/2} c_T(x, -x, \lambda) \cos x \, dx = 0$$

(recall that the function c_T is defined by (4)). This means that the component, orthogonal to the current velocity, of the force acting on a smooth (without roughness) disc is zero. So we obtain $\theta'(\tau) = -\varepsilon R_T(\lambda(\tau))$ where

$$R_T(\lambda) = \frac{1}{2} \int_{[x_0, \pi/2] \cap J} (c_T(x, x, \lambda) - c_T(x, -x, \lambda)) \cos x \, dx, \qquad (13)$$

with $c_T(x, x, \lambda) - c_T(x, -x, \lambda) =$

$$\frac{3 \sin x}{\sin \zeta} \{(\lambda^3 \sin^3 x + 3\lambda \sin x \sin^2 \zeta) \cos \zeta \sin x + (3\lambda^2 \sin^2 x \sin \zeta + \sin^3 \zeta) \sin \zeta \cos x\}$$

and $\zeta = \zeta(x) = \arccos(\lambda \cos x)$. After some algebra we get

$$c_T(x, x, \lambda) - c_T(x, -x, \lambda) = \frac{3 \sin x \cos \zeta}{\lambda \sin \zeta} \{(\lambda^2 - \cos^2 \zeta)^2 + 6 \sin^2 \zeta (\lambda^2 - \cos^2 \zeta) + \sin^4 \zeta\}.$$

Making the change of variable $x \to \zeta$ in the integral (13), we obtain

$$R_T(\lambda) = \int_{[0, \pi/2] \cap \tilde{J}} \frac{3}{2\lambda^3} \{(\lambda^2 - \cos^2 \zeta)^2 + 6 \sin^2 \zeta (\lambda^2 - \cos^2 \zeta) + \sin^4 \zeta\} \cos^2 \zeta \, d\zeta, \qquad (14)$$

where

$$\tilde{J} = \bigcup_{j=0}^{m-1} [\zeta_j, \zeta_j + \Delta\zeta_j],$$

with $\zeta_j = \arccos(\lambda e^{-S_j})$, $\zeta_j + \Delta\zeta_j = \arccos(\lambda e^{-w^{-1}((T_j + \Delta_j)/\varepsilon)})$. Notice that the expression $\{\ldots\}$ in the integral in the right hand side of (14) can be estimated as $\{\ldots\} = \lambda^4 + O(\lambda^3)$ for large values of λ.

Substituting $\lambda = e^{w(\varepsilon, \tau)}$, one obtains

$$\zeta_j = \arccos(e^{w(\varepsilon, \tau) - T_j/\varepsilon}) \quad \text{and} \quad \Delta\zeta_j = \frac{e^{w(\varepsilon, \tau) - T_j/\varepsilon}}{\sqrt{1 - e^{2w(\varepsilon, \tau) - 2T_j/\varepsilon}}} \frac{\Delta_j}{\varepsilon}(1 + o_\varepsilon(1))$$

where $o_\varepsilon(1) \to 0$ as $\varepsilon \to 0$. The value of $\varepsilon R_T(\lambda)$ can now be evaluated as

$$\varepsilon R_T(\lambda) = \varepsilon \frac{3}{2\lambda^3} \left(\lambda^4 \cos^2 \zeta_j \Delta \zeta_j + \hat{R}_j^0(\lambda, \varepsilon) \right) =$$

$$\varepsilon \frac{3\lambda}{2} e^{2w(\varepsilon,\tau)-2T_j/\varepsilon} \frac{e^{w(\varepsilon,\tau)-T_j/\varepsilon}}{\sqrt{1 - e^{2w(\varepsilon,\tau)-2T_j/\varepsilon}}} \frac{\Delta_j}{\varepsilon} + \hat{R}_j^1(\tau,\varepsilon)$$

$$= \frac{3\varphi_j}{2} \frac{e^{4w(\varepsilon,\tau)-4T_j/\varepsilon}}{\sqrt{1 - e^{2w(\varepsilon,\tau)-2T_j/\varepsilon}}} + \hat{R}_j^1(\tau,\varepsilon).$$

Here $|\hat{R}_j^0(\lambda, \varepsilon)| \le C\lambda^3$ where C is a constant; $\hat{R}_j^1(\tau, \varepsilon)$ tends to zero as $\lambda(\tau) \to \infty$, $\varepsilon \to 0$. Thus, we come to the following differential equation for $\theta(\tau)$,

$$\frac{d\theta}{d\tau} = \frac{3\varphi_j}{2} \frac{e^{4w(\varepsilon,\tau)-4T_j/\varepsilon}}{\sqrt{1 - e^{2w(\varepsilon,\tau)-2T_j/\varepsilon}}} + \hat{R}_j^1(\tau,\varepsilon), \qquad \text{if} \quad \tau \in [S_j, S_{j+1} - 1/\sqrt{\varepsilon}].$$

Solutions for this equation are

$$\theta(\tau) = \theta(S_j) + \varphi_j \left[1 - \sqrt{1 - e^{2w(\varepsilon,\tau)-2T_j/\varepsilon}} \left(1 + \frac{1}{2} e^{2w(\varepsilon,\tau)-2T_j/\varepsilon} \right) \right] + \tilde{R}(\varepsilon,\tau), \quad (15)$$

$$\text{if} \quad \tau \in [S_j, S_{j+1}]; \quad j = 0, \ldots, m-1.$$

Here $|\tilde{R}(\varepsilon, \tau)| \le \sqrt{\varepsilon}$ if ε is sufficiently small. The function θ is increasing with respect to τ and with respect to each parameter φ_j. So, we can select all φ_j so that $\theta(S_{j+1}) - \theta(S_j) = \varphi_j^0$. Thus, any part of the trajectory $X([S_j, S_{j+1} - 1/\sqrt{\varepsilon}])$ $(j \ge 1)$ is an arc, close to a line segment of length $(T_{j+1} - T_j)\varepsilon^{-1} - \varepsilon^{-1/2}$.

Let L be the length of the curve g, $\theta_0(\tau)$ be the piecewise constant function, equal to 0 on $[S_0, S_1)$ and equal to $\varphi_1^0 + \ldots + \varphi_{j-1}^0$ on $[S_{j-1}, S_j)$. Then for any $\tau \in [0, L]$ we have

$$\left| X(\tau/\varepsilon) - \frac{1}{\varepsilon} g(\tau) \right| \le \int_{S_0}^{\tau/\varepsilon} |\varphi(s) - \varphi_0(s)| \, ds. \qquad (16)$$

For $\tau \in [\varepsilon S_j, \varepsilon S_{j+1} - \sqrt{\varepsilon}]$ the velocity vector $X'(t/\varepsilon)$ forms an angle $O(e^{-\frac{1}{2\sqrt{\varepsilon}}})$ with the j-th segment of the broken line. This follows from representations (15). On the other hand, contributions of any segment $[\varepsilon S_j - \sqrt{\varepsilon}, \varepsilon S_j]$ to the right hand side of (16) are estimated by $\varepsilon^{-1/2}$. So, we have inequality (12) satisfied if ε is small. $\qquad \square$

9   Conclusion and Discussion

The main results of this paper are the following. Two-dimensional trajectories of bodies, whose boundaries are close to circles, may have (up to rescaling) any shape. The same statement is true for flat curves in the three dimensional real space. Also a description of amphora billiard (quasi-elastic reflector) and its modifications with a wide variety of response functions have been given. All these results are principally novel.

However, our construction while being mathematically correct cannot be implemented in practice. First, we make some non-realistic assumptions that the medium temperature is absolute zero, the particles of the medium do not collide, and (even worse) the collisions of the particles with the boundary of the body are perfectly elastic. Second, even if all these assumptions are satisfied, each cavity should be fabricated with exceptionally high precision, the scale of precision being much smaller than the size of atoms. Third, the path traversed by a disc is proportional to the logarithm of time. Roughly speaking, it may happen that the first meter of the trajectory is traversed in a second, the second meter in a minute, the third meter in a hour, ..., the tenth meter in a billion of years. The experimenter may just not survive the end of the experiment.

Imagine a football player who wants to send the ball so that the trajectory goes round all the players of the rival team and finally gets into the gate. He can indeed do so making use of our results, but the ball surface should be very special; the pressure of the atmosphere should be very low; the Earth gravitation should be negligible; the rival players should be asked not to prevent the (eventually very small) motion of the ball. And it remains to wait. Oh, forgot to say that all this should happen in two dimensions.

Acknowledgements. This work was supported by Russian Foundation for Basic Researches under Grants 14-01-00202-a and 15-01-03797-, by Saint-Petersburg State University under Thematic Plans 6.0.112.2010 and 6.38.223.2014, by FEDER funds through COMPETE – Operational Programme Factors of Competitiveness (Programa Operacional Factores de Competitividade) and by Portuguese funds through the Center for Research and Development in Mathematics and Applications (CIDMA) from the *"Fundação para a Ciência e a Tecnologia"* (FCT), cofinanced by the European Community Fund FEDER/POCTI under FCT research projects (PTDC/MAT/113470/2009 and PEst-C/MAT/UI4106/2011 with COMPETE number FCOMP-01-0124-FEDER-022690). The author is grateful to Prof. Alexandre Plakhov from University of Aveiro for his ideas, remarks and corrections.

References

1. Brock, F., Ferone, V., Kawohl, B.: A symmetry problem in the calculus of variations. Calc. Var. **4**, 593–599 (1996)
2. Bucur, D., Buttazzo, G.: Variational Methods in Shape Optimization Problems, p. 216. Birkhäuser, Boston (2005)
3. Buttazzo, G., Ferone, V., Kawohl, B.: Minimum problems over sets of concave functions and related questions. Math. Nachr. **173**, 71–89 (1995)
4. Buttazzo, G., Kawohl, B.: On Newton's problem of minimal resistance. Math. Intell. **15**, 7–12 (1993)
5. Comte, M., Lachand-Robert, T.: Newton's problem of the body of minimal resistance under a single-impact assumption. Calc. Var. Partial Differ. Equ. **12**, 173–211 (2001)
6. Lachand-Robert, T., Oudet, E.: Minimizing within convex bodies using a convex hull method. SIAM J. Optim. **16**, 368–379 (2006)

7. Lachand-Robert, T., Peletier, M.A.: Newton's problem of the body of minimal resistance in the class of convex developable functions. Math. Nachr. **226**, 153–176 (2001)
8. Lachand-Robert, T., Peletier, M.A.: An example of non-convex minimization and an application to Newton's problem of the body of least resistance. Ann. Inst. H. Poincaré, Anal. Non Lin. **18**, 179–198 (2001)
9. Plakhov, A.: Newton's problem of minimal resistance for bodies containing a half-space. J. Dynam. Control Syst. **10**, 247–251 (2004)
10. Plakhov, A.: Optimal roughening of convex bodies. Canad. J. Math. **64**, 1058–1074 (2012)
11. Aleksenko, A., Plakhov, A.: Bodies of zero resistance and bodies invisible in one direction. Nonlinearity **22**, 1247–1258 (2009)
12. Plakhov A., Exterior Billiards. Systems with Impacts Outside Bounded Domains, p. xiv+284. Springer, New York (2012)
13. Wolf, E., Habashy, T.: Invisible bodies and uniqueness of the inverse scattering problem. J. Modern Optics **40**, 785–792 (1993)
14. Cercignani, C.: Rarified Gas Dynamics. From Basic Concepts to Actual Calculations. Cambridge University Press, Cambridge (2000)
15. Kosuge, S., Aoki, K., Takata, S., Hattori, R., Sakai, D.: Steady flows of a highly rarefied gas induced by nonuniform wall temperature. Phys. Fluids **23**, 030603 (2011)
16. Muntz, E.P.: Rarefied gas dynamics. Annu. Rev. Fluid Mech. **21**, 387–422 (1989)
17. Rudyak, V.Y.: Derivation of equations of motion of a slightly rarefied gas around highly heated bodies from Boltzmann's equation. J. Appl. Mech. Tech. Phy. **14**(5), 646–649 (1973)
18. Bunimovich, A.I.: Relations between the forces on a body moving in a rarefied gas in a light flux and in a hypersonic Newtonian stream. Fluid Dyn. **8**(4), 584–589 (1973)
19. Bunimovich, A.I., Kuz'menko, V.I.: Aerodynamic and thermal characteristics of three-dimensional star-shaped bodies in a rarefied gas. Fluid Dyn. **18**(4), 652–654 (1983)
20. Ivanov, S.G., Yanshin, A.M.: Forces and moments acting on bodies rotating around a symmetry axis in a free molecular flow. Fluid Dyn. **15**, 449–453 (1980)
21. Weidman, P.D., Herczynski, A.: On the inverse Magnus effect in free molecular flow. Phys. Fluids **16**, L9–L12 (2004)
22. Borg, K.I., Söderholm, L.H.: Orbital effects of the Magnus force on a spinning spherical satellite in a rarefied atmosphere. Eur. J. Mech. B/Fluids **27**, 623–631 (2008)
23. Borg, K.I., Söderholm, L.H., Essénm, H.: Force on a spinning sphere moving in a rarefied gas. Phys. Fluids **15**, 736–741 (2003)
24. Plakhov, A.: Newton's problem of the body of minimum mean resistance. Sbornik Math. **195**, 1017–1037 (2004)
25. Plakhov, A.: Billiards and two-dimensional problems of optimal resistance. Arch. Ration. Mech. Anal. **194**, 349–382 (2009)
26. Plakhov, A.: Billiard scattering on rough sets: Two-dimensional case. SIAM J. Math. Anal. **40**, 2155–2178 (2009)
27. Plakhov, A., Tchemisova, T., Gouveia, P.: Spinning rough disk moving in a rarefied medium. Proc. R. Soc. A. **466**, 2033–2055 (2010)
28. Bunimovich, L.A.: Mushrooms and other billiards with divided phase space. Chaos **11**, 802–808 (2001)
29. Porter, M.A., Lansel, S.: Mushroom Billiards. Not. AMS **53**, 334–337 (2006)

Comparative Study on Efficiency
of Mirror Retroreflectors

João Pedro Cruz$^{(\boxtimes)}$ and Alexander Plakhov

Universidade de Aveiro, Aveiro, Portugal
{pedrocruz,plakhov}@ua.pt

Abstract. Here we study retroreflectors based on specular reflections. Two kinds of asymptotically perfect specular retroreflectors in two dimensions, Notched angle and Tube, are known at present. We conduct comparative study of their efficiency, assuming that the reflection coefficient is slightly less than 1. We also compare their efficiency with the one of the retroreflector Square corner (the 2D analogue of the well-known and widely used Cube corner). The study is partly analytic and partly uses numerical simulations. We conclude that the retro-reflectivity ratio of Notched angle is normally much greater than those of Tube and the Square corner. Additionally, simple Notched angle shapes are constructed, whose efficiency is significantly higher than that of the Square corner.

Keywords: Retroreflectors · Geometric optics · Shape optimization · Billiards

Mathematics subject classifications: 49Q10, 49M25, 78A05.

1 Introduction

A retroreflector is an optical device that reverts the direction of incident beams of light [9]. In the framework of geometric optics one deals with light rays that propagate along straight lines (in a homogeneous space) or curves. The retroreflector is called to be *perfect*, if each incident light ray, as a result of interaction with the device, changes its direction to the opposite (such rays are called *retroreflecting*).

Retroreflectors are widely used in economy, for example in road safety and space exploration. Most retroreflectors used in practice are not perfect: only a part of incident light rays are retro-reflected. The most used types of retroreflectors are called *cube corner* and *cat's eye* (see Fig. 1). The former one is based on reflection from three mutually perpendicular planes, and the latter one, on refraction in a lens (and possibly also reflection). Both are not perfect.

A well-known example of perfect retroreflector is *Eaton lens*, a transparent ball with the varying refractive index [2–4,8]. More precisely, the refractive index at a point of the ball equals $n(r) = \sqrt{2R/r - 1}$, where R denotes the radius of the ball and r the distance from the point to the ball center. That is, the index equals 1 at the boundary of the ball and goes to infinity when the point approaches the ball center. An incident light ray passes through the ball and then goes back in the direction opposite to the original one (see Fig. 2).

© Springer International Publishing Switzerland 2015
A. Plakhov et al. (Eds.): EmC-ONS 2014, CCIS 499, pp. 20–32, 2015.
DOI: 10.1007/978-3-319-20352-2_2

Fig. 1. The retroreflectors (a) Cube corner and (b) Cat's eye.

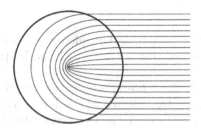

Fig. 2. Eaton lens.

However, design of media with varying refractive index is not an easy task. It seems to be much easier from technical viewpoint to design retroreflectors based solely on mirror reflections, or billiard retroreflectors. Thus, one comes to the problem of creating a perfect billiard retroreflector.

This problem is not solved until now. It is even not known if such retroreflectors really exist. What we know, however, is that there exist *asymptotically perfect* families of retroreflectors [1,5,7]. For arbitrarily small $\varepsilon > 0$, one can choose a retroreflector in such a family so that the portion of light rays reflected from it in wrong directions is smaller than ε. At present, 2D asymptotically perfect retroreflectors are studied in some detail, and almost nothing is known about 3D ones. Notice that 2-dimensional devices may be of interest for practice, especially if the light is supposed to propagate in a single plane. For instance, one can imagine applications in road engineering, when the retroreflectors are placed on the height corresponding to the level of the driver's eyes.

In this paper we concentrate on 2-dimensional asymptotically perfect families of retroreflectors. By a retroreflector we mean a bounded domain B with a marked part of the boundary. The marked part is a line segment (the dashed line in Fig. 3); it is called the *inlet* of the retroreflector. The retroreflector lies on one side of the dashed line, and its boundary is a piecewise smooth curve with finite length.

The propagation of light is represented by the billiard in B. A light ray comes through the inlet, makes several reflections from the boundary ∂B, and finally leaves B through the inlet. The ray is retro-reflected, if the direction of coming in is opposite to the direction of going away ($\varphi = \varphi^+$ in Fig. 3). We

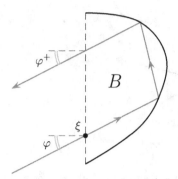

Fig. 3. Motion of light in a retroreflector.

parameterize the inlet by $\xi \in [0, 1]$; each incident ray is naturally labeled by the angle of incidence φ varying from $-\varphi/2$ to $\varphi/2$ and by the point ξ where the ray intersects the inlet.

For a retroreflector B, the *retro-reflectivity ratio* $r(B)$ is the portion of the incident light rays that are retro-reflected by the device. The ratio varies between 0 (no retro-reflection at all) and 1 (perfect retro-reflection). The amount of incoming light is counted according to the natural billiard measure $d\mu(\varphi, \xi) = \frac{1}{2}\cos\varphi\,d\varphi\,d\xi$ defined on $[-\pi/2, \pi/2] \times [0, 1]$. It is a probability measure, that is, $\mu([-\pi/2, \pi/2] \times [0, 1]) = 1$.

It is instructive to calculate the retro-reflectivity ratio of the *square corner*, the 2-dimensional analogue of cube corner (see Fig. 4). Assume that the sides of the corner are perfectly reflecting. Any light ray that makes 2 reflections from the corner reverts its direction, and a ray that makes only 1 reflection goes away in a wrong direction. A simple geometrical analysis allows one to calculate the portion of rays that make double reflections (and therefore are retro-reflected); see Fig. 4 for a graphical illustration. We find that for a fixed φ the portion of retro-reflected rays equals

$$R_{cq}(\varphi) = \begin{cases} 1 - |\tan\varphi|, & \text{if } |\varphi| < \pi/4; \\ 0, & \text{if } |\varphi| \geq \pi/4 \end{cases}$$

(and hence the portion of wrongly reflected rays is $1 - R_{cq}(\varphi)$). Therefore the retro-reflectivity ratio of the square corner equals

$$r_{sq} = \int_{-\pi/2}^{\pi/2} R_{cq}(\varphi) \frac{1}{2}\cos\varphi\,d\varphi = \int_0^{\pi/4} (\cos\varphi - \sin\varphi)\,d\varphi = \sqrt{2} - 1 \approx 0.414.$$

Note that this analysis is not applicable to the second most popular type of retroreflectors: cat's eye. The point is that generally the direction of reversed light rays in cat's eyes does not precisely coincide with the direction of incidence, but is slightly deviated. Thus the formal calculation of the retro-reflectivity ratio would give zero.

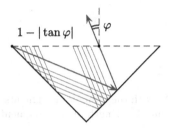

$1 - |\tan \varphi|$

Fig. 4. Reflection of light in a square corner.

At present two asymptotically perfect retroreflectors, *Tube* and *Notched angle*, are known. They are obtained by modifying a rectangle and an isosceles triangle, respectively. Notice that both a rectangle with small $q = (height)/(width)$ and an isosceles triangle with small $q = (base)/(height)$ can serve as retroreflectors (see Fig. 5), and their retro-reflectivity ratios go to $1/2$ as $q \to 0$ (see [6], Chap. 9).

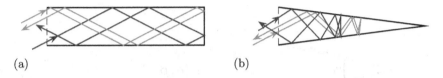

(a)   (b)

Fig. 5. (a) Rectangle-shaped and (b) triangle-shaped retroreflectors. In both cases the retro-reflected ray is shown red, and the wrongly reflected ray is shown blue.

Tube is obtained by removing small periodically located segments parallel to the inlet from a rectangle (see Fig. 6). More precisely, a Tube $B(n, d, \varepsilon)$ is the rectangle $[0, (n+1)d] \times [0, 1]$ with the segments $\{di\} \times [0, \varepsilon]$ and $\{di\} \times [1-\varepsilon, 1]$, $i = 1, \ldots, n$ removed; that is,

$$B(n, d, \varepsilon) = [0, (n+1)d] \times [0, 1] \setminus \cup_{i=1}^{n}(\{di\} \times ([0, \varepsilon] \cup [1 - \varepsilon, 1])).$$

Here the inlet is $\{0\} \times [0, 1]$. Thus, the retroreflector Tube depends on three parameters: n, d, and ε. It is proved in [1,5] that for a certain family of retroreflectors $B(n, d, \varepsilon)$ with $\varepsilon \to 0$, $n = n(\varepsilon) \to \infty$, and with d fixed, the retro-reflectivity ratio $r(B(n(\varepsilon), d, \varepsilon))$ goes to 1 as $\varepsilon \to 0$.

Notched angle is obtained by replacing the lateral sides of the rectangle with a broken line whose segments are parallel and perpendicular to the inlet (see Fig. 7). More precisely, take positive δ, α, β, consider the broken line composed of alternating horizontal and vertical segments inscribed in the angle $x \tan \alpha \leq y \leq x \tan(\alpha + \beta)$ and situated between the vertical lines $x = \delta$ and $x = 1$. Further consider another broken line symmetric to the original one with respect to the x-axis. The initial endpoints of the former and latter lines are $(1, \tan \alpha)$ and $(1, -\tan \alpha)$, respectively.

Fig. 6. The Tube $B(5, 1, 0.12)$, with the number of segments $n = 5$, horizontal distance between the segments $d = 1$, and the length of each segment $\varepsilon = 0.12$.

The Notched angle $B(\alpha, \beta, \delta)$ is bounded by these broken lines and by the vertical lines $x = \delta$ and $x = 1$. The inlet is the vertical segment $\{1\} \times [-\tan \alpha, \tan \alpha]$. Thus, the retroreflector Notched angle depends on three parameters: α, β, and δ. It is proved in [5] that for a certain family of retroreflectors $B(\alpha, \beta, \delta)$ with $\delta = \delta(\alpha) \xrightarrow[\alpha \to 0]{} 0$, $\beta = \beta(\alpha)$, $\beta/\alpha \xrightarrow[\alpha \to 0]{} 0$ the retro-reflectivity ratio $r(B(\alpha, \beta(\alpha), \delta(\alpha)))$ goes to 1 as $\alpha \to 0$.

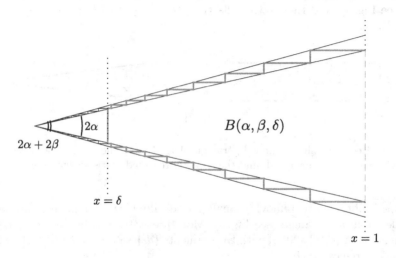

Fig. 7. Notched angle.

Our aim in this paper is to evaluate and compare the efficiency of these retroreflectors. It is supposed that a part of the light is lost after each reflection: a portion k of the light is reflected according to the billiard law, and the portion $1 - k$ is absorbed by the device or scattered. Here $0 < k < 1$. We are going to evaluate the three parameters (a, b, ε in the former case and α, β, δ in the latter case) that provide the maximal, or nearly maximal, retro-reflectivity ratio. Obviously, when k goes to 1 (full reflection), the maximum retro-reflectivity ratio goes to 1 in both cases. We will see, as a result of our study, that for each k the retroreflector Notched angle is much more efficient than the Tube.

Another question we address here concerns creating a reflecting curve with relatively simple shape and with the retro-reflectivity ratio significantly greater than that of the square corner. We will see that such shapes do exist.

Finding a 3D analogue of the Tube and (especially) the Notched angle are challenging tasks for the future.

2 Notched Angle

2.1 Analytical Study

One of the advantages of Notched angle is that its efficiency can be evaluated analytically. Here we provide analytical derivation of the retro-reflectivity ratio $r(\alpha, k)$ in the limit when $\beta \to 0$, $\delta \to 0$. It can be made rigorous with using methods from [5]. However, here we limit ourselves by numerical verification of our heuristics.

First consider two flows of particles with the angles of incidence φ and $-\varphi$, with $0 < \varphi < \alpha$. Fix an indentation (the line ABC in Fig. 8) of our angle and consider the part of the flow incident on it with the angle φ. The segment AB is vertical and BC is horizontal. By unfolding the right triangle ABC one obtains the isosceles triangle AEC. The point F on the side AC is chosen so that the line EF forms the angle φ with EC.

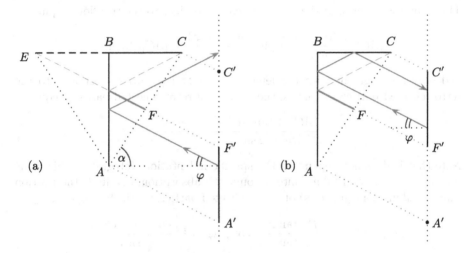

Fig. 8. Particles reflected from an indentation when $|\varphi| \leq \alpha$. (a) A wrongly reflected particle. (b) A retro-reflected particle.

If a particle comes through the segment AF, it makes a reflection and is reflected in a wrong direction (see Fig. 8(a)). If it comes through FC, it makes two reflections and is retro-reflected (see Fig. 8(b)).

The triangle AEF has the angles $\pi - 2\alpha$, $\alpha - \varphi$, $\alpha + \varphi$, respectively. The triangle CEF has the angles α, φ, $\pi - \alpha - \varphi$, respectively. Applying the sine law to these triangles, after a simple trigonometry one obtains

$$\frac{|FC|}{|AC|} = \frac{2\tan\varphi}{\tan\alpha + \tan\varphi}. \tag{1}$$

Consider a vertical line (for example, the inlet of the notched angle). The particles incident on the given indentation with the angle of incidence φ intersect this line at points of a certain segment (the segment $A'C'$ in Fig. 8 (a), (b)). Using again the sine law, one easily finds the length of the segment,

$$|A'C'| = |AC| \frac{\sin(\alpha + \varphi)}{\cos \varphi}.$$

The segment $F'C'$ (see Fig. 8 (b)) corresponds to the retro-reflected particles. Using (1), one easily finds its length,

$$|F'C'| = |AC| \cdot 2 \tan \varphi \cos \alpha.$$

On the other hand, all the particles with the angle of incidence $-\varphi$ (not shown in the figure) are reflected from the indentation in a wrong direction. The length of the corresponding vertical segment equals

$$|AC| \frac{\sin(\alpha - \varphi)}{\cos \varphi}.$$

Thus, the total length of the segments corresponding to the two flows equals

$$|AC| \frac{\sin(\alpha + \varphi)}{\cos \varphi} + |AC| \frac{\sin(\alpha - \varphi)}{\cos \varphi} = |AC| \cdot 2 \sin \alpha,$$

and only one segment with the length $|AC| \cdot 2 \tan \varphi \cos \alpha$ corresponds to the retro-reflected particles. That is, the portion of retro-reflected particles equals

$$\frac{|AC| \cdot 2 \tan \varphi \cos \alpha}{|AC| \cdot 2 \sin \alpha} = \frac{\tan \varphi}{\tan \alpha}.$$

Note that it does not depend on the specific indentation. Integrating this value over $\varphi \in [-\alpha, \alpha]$ and taking into account the absorption, one finds the portion (among all incident particles) of retro-reflected particles with the angles $|\varphi| \leq \alpha$,

$$r_1(\alpha, k) = k^2 \int_{-\alpha}^{\alpha} \frac{\tan \varphi}{\tan \alpha} \frac{1}{2} \cos \varphi \, d\varphi = k^2 \frac{\cos \alpha (1 - \cos \alpha)}{\sin \alpha}.$$

Let now $\varphi > \alpha$. Consider the flow of particles incident on a fixed indentation (the line ABC in Fig. 9) with the angle of incidence φ. Let the triangle EBC be symmetric to the triangle ABC with respect to the line BC. Take the point F on AC so that EF forms the angle φ with BC. If the particle comes through the segment AF, it makes two reflections and goes back in the opposite direction. If it comes through FC then after a single reflection it continues moving forward.

The triangle AEF has the angles $\pi/2 - \alpha$, $\pi/2 - \varphi$, $\alpha + \varphi$, respectively. The triangle CEF has the angles 2α, $\varphi - \alpha$, $\pi - \alpha - \varphi$, respectively. Applying the sine rule to these triangles, one comes to the relation

$$\frac{|FC|}{|AC|} = \frac{\tan \varphi - \tan \alpha}{\tan \varphi + \tan \alpha} =: \lambda_\alpha(\varphi).$$

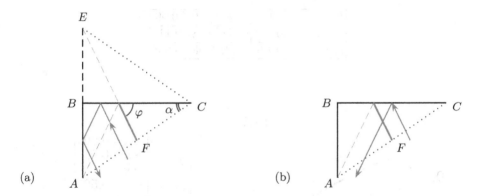

Fig. 9. Particles reflected from an indentation when $\varphi > \alpha$. (a) A retro-reflected particle. (b) A wrongly reflected particle.

This ratio is the portion of the flow that continues moving forward after being reflected in the indentation. It does not depend on the specific indentation.

The process can be described as follows. The part $1 - \lambda$ (where $\lambda = \lambda_\alpha(\varphi)$) of the flow is retro-reflected by the first indentation (after hitting it two times). Thus, the retro-reflected part of the flow is $k^2(1 - \lambda)$. Notice that the first indentation may be different for different particles.

The part of the flow retro-reflected by the second indentation is $\lambda(1-\lambda)$. The corresponding particles make one reflection while moving forward, two reflections in the indentation, and one reflection on the way back — 4 reflections in the total. The retro-reflected part of the flow is $k^4\lambda(1 - \lambda)$.

Continuing this process, one obtains the sequence $k^{2m+2}\lambda^m(1 - \lambda)$, $m = 0, 1, 2, \ldots$, the sum of its terms being the portion of retro-reflected particles,

$$k^2(1 - \lambda)[1 + k^2\lambda + k^4\lambda^2 + \ldots] = \frac{k^2(1 - \lambda)}{1 - k^2\lambda}.$$

Here $\lambda = \lambda_\alpha(\varphi)$. Notice that this portion is related to the flow with the angle of incidence φ. The portion of retro-reflected particles corresponding to all angles $\varphi > \alpha$ is obtained by integration,

$$r_2(\alpha, k) = \int_\alpha^{\pi/2} \frac{k^2(1 - \lambda_\alpha(\varphi))}{1 - k^2\lambda_\alpha(\varphi)} \cos\varphi \, d\varphi = \int_{\tan\alpha}^\infty \frac{2k^2}{\frac{t}{\tan\alpha}(1 - k^2) + (1 + k^2)} \frac{dt}{(1 + t^2)^{3/2}}.$$

Thus, the retro-reflectivity ratio equals

$$r(\alpha, k) = r_1(\alpha, k) + r_2(\alpha, k)$$
$$= k^2 \frac{\cos\alpha(1 - \cos\alpha)}{\sin\alpha} + \int_{\tan\alpha}^\infty \frac{2k^2}{\frac{t}{\tan\alpha}(1 - k^2) + (1 + k^2)} \frac{dt}{(1 + t^2)^{3/2}}. \qquad (2)$$

The graphs of $r(\alpha, k)$ as functions of α for several values of k are shown in Fig. 10. The optimum angle $\alpha_{\max} = \alpha_{\max}(k)$ is indicated by a dot on each curve. The values of the angles $\alpha_{\max}$ and the corresponding retro-reflectivity ratios are given in the table below.

k	0.5	0.8	0.9	0.99
$\alpha_{\max}$	0.445	0.356	0.287	0.117
$r(\alpha_{\max}, k)$	0.146	0.421	0.582	0.864

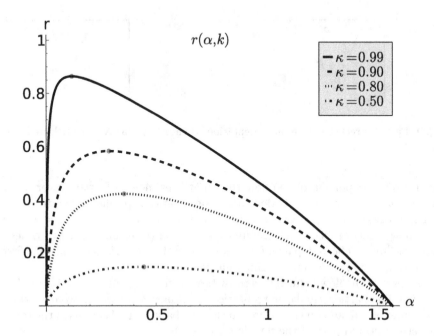

Fig. 10. Retro-reflectivity ratio $r(\alpha, k)$ of Notched angle for several values of k.

2.2   Numerical Simulation

The number of light beams coming through the inlet was chosen to be 1000 in all experiments. If $\bar{r}$ is the estimate of the retro-reflectivity ratio in a simulation, then the standard deviation of the estimate is less than 2 %, as can be seen from the formula for the deviation of Normal distribution $\sqrt{\bar{r}(1 - \bar{r})/1000}$ for $\bar{r} \in [0, 1]$.

In the case of positive β and δ there are no analytic formulas, so one needs to proceed to numerical simulation. First we verify theoretical results for the limiting case $\beta \to 0$, $\delta \to 0$. To that end, we calculate the retro-reflectivity ratio $r(\alpha_{\max}, \beta, \beta, k)$ for the fixed values $k = 0.9$ and $k = 0.99$ and the corresponding optimal values $\alpha_{\max} = \alpha_{\max}(0.9) = 0.287$ and $\alpha_{\max} = \alpha_{\max}(0.99) = 0.117$, and consider the values $\beta = \delta$ varying from 0.01 to 0.1. One sees in Fig. 11 that the retro-reflectivity ratio approaches the corresponding value $r(\alpha_{\max}, k)$ (marked by a point on the vertical axis) as β goes to 0.

The three graphs of maximum retro-reflectivity ratio versus k with fixed values of β and δ are shown in Fig. 12. The two graphs are related to the values

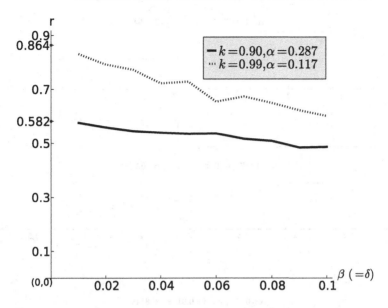

Fig. 11. The plot of the retro-reflectivity ratio *vs* $\beta(=\delta)$ for $k = 0.9$ and $k = 0.99$.

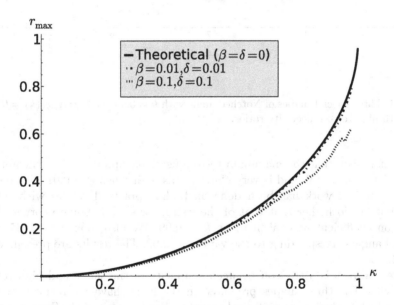

Fig. 12. The graphs of maximum retro-reflectivity ratio *vs* k in the three cases when (a) $\beta = \delta = 0.1$; (b) $\beta = \delta = 0.01$; and (c) the theoretical limiting case for $\beta = \delta = 0$.

$\beta = \delta = 0.1$ and $\beta = \delta = 0.01$, and the third graph plots the theoretical maximum value $\max_\alpha r(\alpha, k)$, which corresponds to the limiting case $\beta = \delta = 0$.

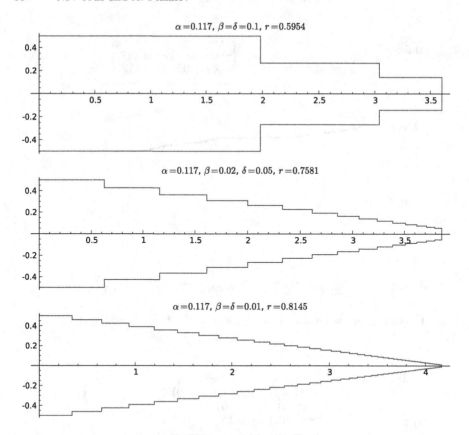

Fig. 13. Three special shapes of Notched angle with $k = 0.99$ and $\alpha = \alpha_{\max}(k) = 0.117$ and with high retro-reflectivity ratios.

Again, it is seen that the maximum retro-reflectivity approaches its theoretical limit as $\beta \to 0$, $\delta \to 0$, and is very close to this limit when $\beta = 0.01$, $\delta = 0.01$.

A numerical work has been done on finding practical shapes with retro-reflectivity ratio higher that that of the well-known shape Square corner. The reflection coefficient was taken to be $k = 0.99$. We also took $\alpha = 0.117$, the optimal angle corresponding to the value $k = 0.99$. The results are presented in Fig. 13.

There always is a tradeoff between simplicity of the shape and high retro-reflectivity. The three shapes presented in the figure have the ratios $r \approx 0.6$, 0.76, and 0.82, which are significantly greater than the ratio of the Square corner $k^2(\sqrt{2} - 1) \approx 0.4$. Naturally, the greater the ratio, the more complicated is the shape: in the first case one has $\beta = \delta = 0.1$, in the second case $\beta = 0.02$, $\delta = 0.05$, and in the third case $\beta = \delta = 0.01$.

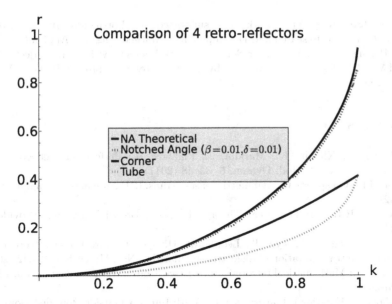

Fig. 14. The plots of the retro-reflectivity ratio vs k for the retroreflectors (a) Tube with $n = 49$, $d = 0.2$, $\varepsilon = 0.05$, (b) Notched angle with $\beta = \delta = 0.01$ and with $\alpha = \alpha_{\max}(k)$, (c) the square corner, and (d) $r(\alpha_{\max}, k)$, the retro-reflectivity ratio of Notched angle with $\beta \to 0$, $\delta \to 0$.

3   Tube and Comparison with Notched Angle

It was proved in [1,5] that there exist Tube retroreflectors with retro-reflectivity ratio arbitrarily close to 1. More precisely, suppose that the sides of the retrore-flector are perfectly reflecting, $k = 1$; then we have $r(B(n,d,\varepsilon)) \xrightarrow[\varepsilon \to 0]{} 1$ for a certain family of retroreflectors with the size of small segments going to zero, $\varepsilon \to 0$, and their number going to infinity, $n = n(\varepsilon) \to \infty$, and with fixed distance d between the segments.

There are no analytical formulas for the retro-reflectivity ratio of Tube with positive values of the parameters ϵ, n, and d, so we did an extensive numeri-cal simulation with ϵ taking the values in $\{0.01, 0.05, 0.1, 0.2\}$ and with $k < 1$, searching for the values of n and d that provide the best retro-reflectivity ratio. We found that the retro-reflectivity ratio of Tube is generally much smaller than that of Notched angle, and even of the square corner, as seen in Fig. 14. The reason is that Tube requires a huge number of light reflections. As a result, when $k < 1$, a large portion of light is absorbed, thus lowering the retro-reflectivity ratio.

In Fig. 14 the retro-reflectivity ratios of three retroreflectors are compared: (a) Tube with $n = 49$, $d = 0.2$, $\varepsilon = 0.05$; (b) Notched angle with $\beta = \delta = 0.01$ and with $\alpha = \alpha_{\max}(k)$ taken to be optimal, (c) the square corner, and (d) the function $r(\alpha, k)$ (2) with $\alpha = \alpha_{\max}(k)$ (recall that it defines the retro-reflectivity ratio in the limiting case of Notched angle with $\beta \to 0$, $\delta \to 0$).

Acknowledgements. This work was supported by Portuguese funds through CIDMA– Center for Research and Development in Mathematics and Applications and FCT – Portuguese Foundation for Science and Technology, within the project PEst-OE/ MAT/ UI4106/ 2014, as well as by the FCT research project PTDC/ MAT/ 113470/ 2009.

References

1. Bachurin, P., Khanin, K., Marklof, J., Plakhov, A.: Perfect retroreflectors and billiard dynamics. J. Mod. Dynam. **5**, 33–48 (2011)
2. Eaton, J.E.: On spherically symmetric lenses. Trans. IRE Antennas Propag. **4**, 66–71 (1952)
3. Luneburg, R.K.: Mathematical Theory of Optics. Brown University, Providence (1944)
4. Ma, Y.G., Ong, C.K., Tyc, T., Leonhardt, U.: An omnidirectional retroreflector based on the transmutation of dielectric singularities. Nat. Mater. **8**, 639–642 (2009)
5. Plakhov, A.: Mathematical retroreflectors. Discr. Contin. Dynam. Syst.-A **30**, 1211–1235 (2011)
6. Plakhov, A.: Exterior Billiards: Systems With Impacts Outside Bounded Domains, XIII, 284 pp., 108 illus. Springer, New York (2012)
7. Plakhov, A., Gouveia, P.: Problems of maximal mean resistance on the plane. Nonlinearity **20**, 2271–2287 (2007)
8. Tyc, T., Leonhardt, U.: Transmutation of singularities in optical instruments. New J. Phys. 10(115038), 8pp (2008)
9. Walker, J.: Wonders with the retroreflector. The Amateur Scientist. Scientific American, April 1986

Optimization and Applications

Multicriteria Optimization in a Typical Multi-Isle Warehouse with Multiple Racks

Diana G. Ramirez-Rios[1](✉), Laura P. Manotas Romero[1],
and Jairo R. Montoya-Torres[2]

[1] Fundación Centro de Investigacion en Modelacion Empresarial
del Caribe (FCIMEC), Carrera 60 No. 64-122, Barranquilla, Colombia
{dramirez,lpmanotas}@fcimec.org
[2] Escuela Internacional de Ciencias Económicas y Administrativas,
Universidad de La Sabana, Km 7 Autopista Norte de Bogotá D.C.,
Chía, Cundinamarca, Colombia
jairo.montoya@unisabana.edu.co

Abstract. This paper considers two common problems frequently found in warehouses: slotting and picking. The former refers to the best arrangement of items in the warehouse, while the latter concerns the definition of the best route to pick up the selected objects. In most industrial practice, the implementation of picking and slotting optimization techniques uses information based on historical data that, in most cases, would fail to work because of many factors affecting daily operations in the warehouse. Simulation models have been employed to build virtual scenarios in order to predict the outcomes of a specific operational decision. Simulation models also fail because collected data is not fully reliable. In order to overcome those problems, this paper proposes the use of a hybrid simulation and optimization approach in which real-time data is incorporated thanks to radio-frequency identification (RFID) technology. Operational decisions are hence made in real-time. The approach is validated using real data from a pharmaceutical manufacturer.

Keywords: Warehousing · Multi-criteria optimization · Simulation · Information processing · RFID

1 Introduction

Because of globalization of marketplaces, enterprises have completely reconfigured their supply chains in order to increase customer service levels and respond to demand variability. Warehouses play a pivotal role in the supply chain [1] and requirements for warehousing operations have significantly increased. Specifically, the customer needs in terms of order accuracy and response time, order frequency, order quantity and order size have dramatically changed with the globalized economy and new demand behavior. The academic literature has widely debated the issues of warehouse design and management, mainly focusing

© Springer International Publishing Switzerland 2015
A. Plakhov et al. (Eds.): EmC-ONS 2014, CCIS 499, pp. 35–48, 2015.
DOI: 10.1007/978-3-319-20352-2_3

on minimizing operational costs and delivery times, contributing to supply chain performance. The interested reader can refer to the comprehensive surveys on warehouse and industrial storage system issues proposed in [2–4].

Warehousing problems have been considered a critical issue in the supply chain. Its importance is mainly given by the fact that the efficiency of the entire supply chain depends on how inventory is managed and stored [5]. In order to guarantee this efficiency, it is necessary to optimize both inventory management and storage. Management problems in warehouses includes stock traceability as one of the most important issues. Most of the problems of inventory management are based on where and how the product is stored. Storage is becoming increasingly expensive and customers are demanding higher product rotation, thus minimizing warehouse space and speeding purchasing and supplying processes. However, this processes must be very effective and accurate so that it does not produce unnecessary costs. Another issue to address in warehouse management concerns the picking operation of orders. Picking is known as the process of handling inventory inside the warehouse, involving the pick-up of inventory that arrives to the warehouse and transportation to the corresponding racks or shelves, the pick-up from these racks or shelves and transportation to production or delivery areas.

This paper studies those two problems: inventory storage and picking. We consider a classical configuration of a warehouse with multiple isle and multiple racks, organized in blocks, containing slots with uniform capacities. For an optimal performance of the warehouse, a multicriteria optimization problem is solved where criteria considered are [6]: (1) total occupation of the warehouse, (2) total picking distances and (3) average stock rotation. To this end, our solution approach hybridizes classical multicriteria models with computer simulation and integrates radio-frequency identification (RFID) tools for better information management. RFID, one of the Automatic Identification and Data Capture (AIDC) technologies, is employed as information management tool. It has attracted significant attention in the fields of supply chain and manufacturing, and more recently, in various service sectors thanks to the advantages offered over other AIDC technologies such as barcodes [7,8]. The benefits of information tracking and tracing through RFID are well-perceived by industry, including retail, logistics, manufacturing, military, healthcare, pharmaceuticals and the service sector [9].

This paper is organized as follows. The problem under study is explained in detail in Sect. 2, while the review of related literature is given in Sect. 3. Section 4 presents the solution approach based on simulation, optimization and the implementation of RFID for data accuracy. The numerical validation inspired from a real life case is presented in Sect. 5. Finally, some conclusions and opportunities for further research are drawn in Sect. 6.

2   Problem Description

We consider a warehouse composed of multiple isles and racks with fixed capacities. Each rack has multiple levels and in each level, several slots in depth. Each slot has a capacity of a Standard American Pallet (1×1.2 m), as shown in Fig. 1. Several authors have approached this type of warehouse [10–14]. Some specific considerations are given to this problem:

- This warehouse only manages pallets of different families of products that may have some location constraints given their special features.
- Products are only managed in pallets using a randomized policy. It is to note, however, that a dedicated policy can easily be implemented in our proposed approach.
- Several zones are considered in the warehouse: approved (ap), quarantine (qt), rejected (r), retained (rt), returned (d) and picking (a) zones.
- The same velocity is considered by the vehicles that operate the picking process.
- A small percentage of product located in the quarantine zone are finally rejected or retained (this is defined by the experienced manager). The remainder set of products goes to approved zone.
- Distances between positions in the warehouse are determined from the entrance to the central point in each rack, between racks, and from each rack to the exit zone, thus, obtaining a network as shown in the figure. If there is more than one route to the same position, the shortest path is considered.

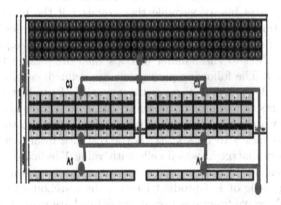

Fig. 1. Global warehouse network for calculating distances

Studies developed in warehousing optimization indicate that the most critical problem concerns the picking operation. Thus, minimizing the total distance traveled by the vehicle is one of the main objectives studied in warehousing. This problem is approached as the TSP (Traveling Salesman Problem) and many solution approaches have been given through metaheuristics (e.g. [15–20]). These approaches give good results, yet, they were not applicable to real industrial cases. In real practice, simple heuristics have been applied in order to minimize the distance traveled but they do not consider other aspects in the order routes.

Another objective functios considered in picking optimization is the minimization of order cycle time, as it is a measure of the rotation of the product and its location in the warehouse.

Product arrangement in the specific locations, is another important problem considered in warehouses. It is known as slotting optimization. Several objectives may be considered: throughput maximization, space utilization minimization or storage costs minimization. An important contribution to the slotting problem was the Cube Per Order (CPO) rule [21]: an algorithm developed to minimize the costs involved in the waiting time for the selection of products in a given order. The application of CPO to optimally assign the picking area in a warehouse was given by [22]. All the objectives mentioned above are important and each one of them has been applied separately. The picking distances may be minimized by one of the algorithms mentioned above, yet, it may be possible that the throughput of the products with this algorithm decreased and is not maximized as it should be, because some orders are priority and need to be considered for its high rotation and special characteristics.

For an optimal performance of this warehouse, a multicriteria optimization problem is considered, where the criteria considered are the following: (1) total occupation of the warehouse, (2) total picking distances and (3) average inventory stored at the warehouse. As a result, this multi-criteria optimization model aims for the maximization of the first objective and the minimization of the other two objects, as shown in the following mathematical formulation. Decision variables are: y_{ijt} as a binary variable that equals 1 if the product is located in the rack j at time t, or 0 otherwise; x_{it} as a binary variables that equals 1 if the product is picked at the time t, and 0 otherwise; and u_{it} as an integer variable corresponding to the number of trips that the picking team picks up product i at time t. The following parameters are defined: cap_{jt} corresponds to the capacity of rack j at time t; q_{it} is the quantity of product i requested at time t; v_i represents the volume of product i (in m^3); V_{total} is the total volume managed by the warehouse; de_j is the distance from entrance to rack j; d_{jk} is the distance from rack j to rack k; ds_j is the distance from rack j to exit; e_{it} represents a binary matrix in which cells with value 1 indicating that product i enters the warehouse at time t, and 0 otherwise; and s_{it} is a binary matrix with cells taking value of 1 if product i leaves the warehouse at time t, and 0 otherwise. The objective functions previously defined are respectively expressed in mathematical form as follows:

2.1 Objective Functions

$$Zmin = \sum_t \sum_j \sum_i \frac{q_{it} v_i y_{ijt}}{V_{total}}$$

$$Zmin = \sum_t \sum_j \sum_i u_{it} de_j e_{it} y_{ijt} + \sum_t \sum_j \sum_i u_{it} x_{it} d_{jk} y_{ijt} + \sum_t \sum_j \sum_i u_{it} x_{it} ds_j s_{it} y_{ijt}$$

$$Zmin = \sum_t \sum_j \sum_i q_{it}y_{ijt}$$

Some constraints have to be defined. Constraints (1) indicate that an item is picked from a given location at a given time t if it is actually located at this location. The sets of Constraints (2) and (3) are capacity constraints. Finally, Constraints (4) guarantee that each product is assigned to only one slot.

$$\sum_j cap_{jt}y_{ijt} \geq x_{it}, \forall i, t \tag{1}$$

$$\sum_i q_i x_{it}u_{1j} \leq cap_{jt}, \forall j, t \tag{2}$$

$$\sum_i z_{it}u_{5j} \leq cap_{jt}, \forall j, t \tag{3}$$

$$\sum_i y_{ijt} = 1, \forall j, t \tag{4}$$

3   Review of Related Literature

3.1   Optimization in Warehouses

Most of research in warehousing management is related to the minimization of routing distances inside the warehouse [10,23]. One of the classic mathematical models was developed by [24], based on a stock size and location problem in a multi-dimensional warehouse that integrates inventory and picking costs. The objective was the minimization of inventory storage costs and costs associated to the picking operation, assuming both single and double deep pallet racking. In a typical warehouse, composed of several parallel rows, a central depot and two possibilities to change rows (for the front and rear of the warehouse), several heuristics are known for the optimal routing of the picking operation. The algorithm proposed in [14] has been recognized to be very efficient to find the shortest path for picking. For more complex design, other types of heuristics have been proposed, such as the S-shaped and the Largest Gap Heuristic. Dynamic programming has also been applied [12]. The connection between blocks is made such that the distance traveled is minimized across the rows. This heuristic combines the S-shaped with the Largest Gap. The so called Aisle-by-Aisle (row after row) heuristic was proposed in [13], in which the course is done in all the rows that contain items, starting with the one that contains items. The Pick-Path optimization algorithm is presented in [11], while an optimization approach based on control systems is presented in [10]. Metaheuristics procedures have also been considered, mainly to optimize the picking operation. Example of such procedures are Ant Colony Optimization (ACO), Genetic Algorithms, Simulated Annealing, and Particle Swarm Optimization [25]. For slotting optimization, several heuristics can be found in the literature: Cube Per Order (CPO) [20], Direct Search [26] and Gradient Search [24]. Metaheuristics can also be found. As an example, the item relocation problem is solved in [27] using Tabu Search by considering it as a slotting problem.

In real industrial practice, most companies are looking for a methodology to achieve optimal SKU placement without the need of expensive information systems. [28] studied the effect in terms of travel distance and material handling time reductions, of an optimal rather than a uniform item allocation in one-block picking warehouses, both with and without the use of a simple picking heuristic. It is possible to see that improvements due to a better slot-code optimization are reduced when the heuristic is used. The importance of these results is crucial, especially for manufacturing, distribution and retailing companies seeking an efficient design for their warehouse.

The work carried out in [29] presents an Operations Research-oriented solution to provide a visible reduction of the overall required warehousing space. In addition to develop an effective multi-product slot-code, these authors focused on finding a cost-effective way to solve the storage location assignment problem through a mathematical optimization approach. Results showed that, even using a dedicated storage policy approach, the outcomes obtained with their model reached the lower bound computed using a randomized policy which should be unavoidably sustained by warehouse management system software. Finally, the work in [30] was to minimize the required storage space while finding a good slot-SKUs allocation in order to reduce handling times and distances. The paper presented an original multi-product slot allocation heuristic developed by approaching it as a vertex coloring problem [31,32]. The approach is evaluated on a real industrial case and demonstrated its effectiveness since performances were significantly close to this best conceivable case.

3.2   Multicriteria Optimization to Warehousing Models

Multi-criteria decision making (MCDM) is divided in two categories: multiple attribute decision making (MADM) and multiple objective decision making (MODM) [33]. One example of MADM, and perhaps the most employed, is the Analytic Hierarchy Process (AHP) [34]. In warehouse management, AHP has been employed to determine the relative weightings of alternative warehouses, taking into account criteria such as delivery costs and customer service level. On the other hand, MODM techniques have been employed in the literature. One example is goal programming (GP) [35] in order to incorporate system restrictions, resources and the AHP priority to select the best set of warehouses without exceeding the available resources [33]. The AHP has also been applied to the transshipment problems [36] by including financial and non-financial issues. Nevertheless, some decision making problems cannot be directly solved by applying the AHP technique [36]; fuzziness can be introduced [37]. Another example of application of the AHP method for the operational optimization process in warehouses is presented in [38]. The goal was to reduce product handling costs by minimizing the picking process and locating the products in the correct position.

4 Solution Approach

Considering all the aspects involved in the dynamics of an industrial warehouse store, this paper presents a solution approach for the type of warehouse described above, based on simulation and optimization models. In order to solve the three objective functions, it is assuming that supplying process to the warehouse has been adjusted and improved, so data can be taken in an automatic way through Radio Frequency Identification (RFID) system [39]. Thus, it is considered as a real-data modeling process, as the orders are made daily and those orders can change with the market dynamics.

The methodology applied is an integration of simple heuristic models, mixed-integer linear programming (MILP) optimization and discrete-event simulation modeling in order to obtain the best scenarios of the operation in any warehouse facility with the characteristics mentioned above. This section describes in detail the proposed solution approach.

4.1 Phase 1. Heuristic for an Optimal Configuration of the Families in the Racks

Step 0. Initialize decision vectors and mining on historical data. Set $Counter = 0$
Step 1. Select the picking zone based on historical data.
Step 2. Allocate of the families to the racks using a priority rule based on the rotation of the product, subject to capacity constraints.
Step 3. Evaluate of the objective function based on the total distance traveled:

$$Z = \sum_i \sum_j dis E_j X_{ij} Y_{1j} Dem_i + Q_i d_{jj} X_{ij} Dem_i + Q_i d_{jj} X_{ij} Y_{3j'} Dem_i + dis E_j Y_{3j} Dem_i$$

If $Counter$ is greater than the number of available racks in Est $Alistamiento_j$, **Then**: choose the scenario that minimizes Z, the total distance traveled in the picking operation; **Else**: Save scenario and objective function value, Z; increase $Counter = Counter + 1$, and **go to** Step 2.

4.2 Phase 2. Optimal Allocation of Orders to Positions of the Warehouse

Setup the configuration of the families assigned in Phase 1 to the simulation model. Each order is assigned to the racks corresponding to each family and its position is based on the minimization of the average time required to fulfill all orders, according to the following mathematical model. Parameters are defined to be R_i: annual rotation of item i; B: storage capacity of rack; V is the speed of the vehicle (mt/min). Binary decision variables are: $Pos_{ixyz} = 1$ if item i is located in position (x, y, z), and 0 otherwise. The model is:

$$Zmin = \sum_i \sum_x \sum_y \sum_z \frac{(x + y + z) Pos_{ixyz} R_i}{V}$$

Subject to

$$\sum_x \sum_y \sum_z Pos_{ixyz} = 1, \forall i \in I \tag{5}$$

$$\sum_i Pos_{ixyz} \leq 1, \forall x, y, z \tag{6}$$

$$\sum_i \sum_x \sum_y \sum_z Pos_{ixyz} \leq B \tag{7}$$

$$Pos_{ixyz} \in \{0, 1\}, \forall x, y, z, i \tag{8}$$

Each rack has a subproblem associated to this mathematical model. Optimization in each rack will lead to a scenario becomes an input of the simulation model.

4.3 Phase 3. Generation of Results Based on Each Scenario Generated

Given the results in each scenario given in Phase 1 and obtaining the optimal allocation of items in Phase 2, the discrete-event simulation model generates results that measure: (1) total occupation of the warehouse, (2) total picking distances and (3) average stock rotation. The Pareto Optimal Solutions are selected and the scenarios are proposed for the final user to decide which one to implement in the real-life situation.

5   Numerical Implementation and Analysis of Results

A case study was developed from the cosmetics and pharmaceutical industry. We studied the packing and container warehouse. For the numerical implementation of the proposed approach, historical data was gathered for the last year of operation. This historical data included the material entering and exiting the warehouse, the value of the material managed, and the forecast of the demand for a year of operation. Figure 2 shows the warehouse layout and the network for the calculation of distances. Tables 1, 2, and 3 show the summary of the historical data gathered for the implementation. Table 2 also presents the cost of average inventory $(CEn - CEx)$, which is the cost of product entering minus the cost of product exiting the system.

Products are grouped by families, since the products of the same family are very similar and are used for the same customer. For example, the packaging of B family are only used for B customer orders; F-type customer orders are made with the packaging of families F or Fc, according to the kind of product to be delivered. This division is done to facilitate inventory management, because to there are many types of packaging for the same product. Usually, when orders are placed, several references of the same family are requested, varying amounts, and depending on the season. Inventory replenishment takes place every fortnight. So, the average inventory in the warehouse is the annual total divided by 24 periods (Quart Dem). There is a difference between the maximum inventory

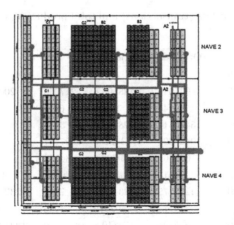

Fig. 2. Warehouse layout with network of rack positions

Table 1. Basic information per family

Id	Name of family	Annual dem	Quart dem	Max inv
1	A	194	8,08	7
2	B	72	3	3
3	Ca	834	34,75	28
4	Cb	1816	75,67	61
5	Cc	2918	121,58	98
6	D	11942	497,58	399
7	Ea	2440	101,67	82
8	Eb	236	9,83	8
9	F	426	17,75	15
10	Ec	594	24,75	20

Table 2. Cost of average inventory per family

Id	Cost of prod entering	Cost of prod exiting	CEn-CEx
1	$ 12.145.435.823	$ 347.309.819	$ 11.798.126.004
2	$ 300.245.396	$ 10.549.921	$ 289.695.475
3	$ 6.718.999.386	$ 39.860.761	$ 6.679.138.626
4	$ 6.848.164.104	$ 201.863.151	$ 6.646.300.953
5	$ 2.465.259.803	$ 200.155.475	$ 2.265.104.328
6	$ 540.337.500	$ 140.647.500	$ 399.690.000
7	$ 3.558.216.498	$ 129.858.197	$ 3.428.358.301
8	$ 3.725.192.948	$ 142.611.620	$ 3.582.581.328
9	$ 3.693.482.868	$ 141.597.863	$ 3.551.885.005
10	$ 11.947.935.083	$ 343.402.902	$ 11.604.532.182

Table 3. Frequency of orders (transactions) and total product managed per family per zone

	Transactions (i, k)						Ordered quantities (i, k)					
Id	1	2	3	4	5	6	1	2	3	4	5	6
1	303	243,4	134	1	0	0	8931946	7188908,8	217948	200	0	0
2	10	36	12	0	0	0	5245	15824	6524	0	0	0
3	582	465,6	14	0	0	0	15410614	12328491,2	3549	0	0	0
4	184	147,2	17	0	0	0	10427695	8342156	27141	0	0	0
5	145	142	207	0	0	0	642661	649728,8	86148	0	0	0
6	99	79,2	101	0	0	0	24015	19212	6251	0	0	0
7	73	193,4	32	1	0	0	4293772	3483401,6	7087	2376	0	0
8	51	52,8	59	1	0	0	370775	407575	4686	40	0	0
9	63	185,4	21	3	0	0	215426	172340,8	3210	2212	0	0
10	103	217,4	54	0	1	0	6149668	4919734,4	49795	0	8100	0
						Total	46.471.817	37.527.372,60	412.339,00	4.828,00	8.100	0
						$P(k)$	55,0454 %	44,4508 %	0,4884 %	0,0057 %	0,0096 %	0,0 %

stored in the warehouse (space restrictions) against the actual average inventory; the warehouse has space problems.

Results generated as part of the implementation of the proposed methodology indicated five Pareto Optimal Solutions. These solutions were evaluated in the three objectives functions mentioned above, as shown in Table 4. From the five scenarios, three were attractive for the operators of the warehouse S0*, S1* and S2*. They were simulated and results were compared to the original scenario, as shown in the three components (each objective function) of Fig. 3. These results indicate the efficiency of the proposed solution method. These results not only indicate a better efficiency of the warehouse, but also allows the decision maker to implement the best decision based on a careful selection of the best scenarios. By having these three scenarios, the decision maker is able to take into account, possible S2 as the best scenario, although it is not the best among all three objectives.

Table 4. Results for all three criteria analyzed per scenario

	Total dist. traveled	W. occupation	Avg. inv. stored
S0	2.758.165	0,3147	384417,089
S1	2.187.116	0,5985	797406,4236
S2	2.867.968	0,5171	379408,0728
S3	2.123.826	0,6287	870084,044
S4	2.602.788	0,4216	554218,125

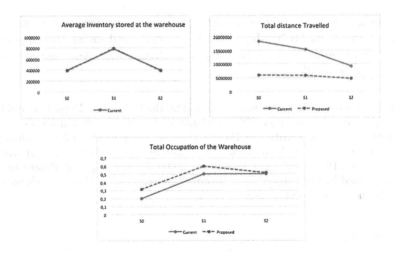

Fig. 3. Results of scenario comparison for each objective function

6   Conclusions and Perspectives

Allocating and distributing the items in a warehouse is not an easy task, especially due to the diversity of the products managed and the characteristics they may have when assigning a position inside the warehouse. Also, due to the large quantities managed, at a daily basis, it is necessary that allocation is done in a fast and efficient manner. Optimization models, on the one hand, offer this efficiency in the allocation of products and the configuration of the warehouse. Simulation models, on the other hand, offer flexibility and visibility in the execution of scenarios, as well as the generation of indicators that need to be taken in consideration when making the best decision.

This paper approached a typical multi-isle warehouse with multiple racks, at multiple levels, where only pallets are managed. This problem searches for the optimal results in three criteria: (1) total occupation of the warehouse, (2) total picking distances and (3) average inventory stored at the warehouse. The solution approach considered was a methodology that used both optimization and simulation methods. Results obtained with the application on a case study from the Cosmetics and Pharmaceutical Industry demonstrated a significant improvement in the configuration of the warehouse. Although some scenarios appeared to be better than others in total distance traveled, others had less occupation and a higher levels of inventory stored. Under multicriteria optimization, the user has several options that are all Pareto Optimal. In addition, performance indicators generated by the simulation model give the user other aspects to consider in the decision making process.

The warehousing optimization problems are very diverse and generalities to this problem can become complex with regards to reaching an optimal approach to solve this type of problems. This is positive because research directions in the area remain open. In the first instance, it is not easily categorized and not easily compared to other solution approaches. Secondly, other configurations of warehouses and material handling should be considered under this approach. Finally, it is interesting to analyze warehouses that manage heterogeneous sizes of products to be handled.

Acknowledgments. This work was supported by the Colombian Department of Science, Technology and Innovation COLCIENCIAS and Centro de Investigación en Modelación Empresarial del Caribe (FCIMEC) through the project entitled: "Diseño e implementaciòn de centros de almacenamiento automatizados mediante la aplicación de tecnologí as EPC-RFID", grant number 2233-454-25947. Special thanks to the team that worked in this Project: Luis Ramirez, Lauren Castro, Miguel Jiménez, Erik Maldonado and Fernando González.

References

1. Accorsi, R., Manzini, R., Maranesi, F.: A decision-support system for the design and management of warehousing systems. Comput. Ind. **65**(1), 175–186 (2014)
2. De Koster, R., Le-Duc, T., Roodbergen, K.J.: Design and control of warehouse order picking: a literature review. Eur. J. Oper. Res. **182**(2), 481–501 (2007)
3. Gu, J., Goetschalckx, M., McGinnis, L.F.: Research on warehouse operation: a comprehensive review. Eur. J. Oper. Res. **177**(1), 1–21 (2007)
4. Dallari, F., Marchet, G., Melancini, M.: Design of order picking system. Int. J. Adv. Manuf. Technol. **42**(1–2), 1–12 (2009)
5. Strack, G., Pochet, Y.: An integrated model for warehouse and inventory planning. Eur. J. Oper. Res. **204**(1), 35–50 (2010)
6. Ramirez, D.G., Ramirez, L.E., Castro, L.J., Jimenez, M.A., Manotas, L.P.: RFID Implementation and simulation-based system dynamics for optimizing warehousing strategies under multiple criteria. In: Proceedings of the Tenth LACCEI Latin American and Caribbean Conference (LACCEI 2012), July 23–27, Panama City, Panama (2012)
7. Baker, P.: Will tags get out into the supply chain? Works Manag. **58**(2), 34–37 (2005)
8. Patadia, S., Dua, S., Meyers, M.: Mike Meyers' CompTIA RFID + Certification Passport. McGraw-Hill Publishing, New York (2007)
9. Ngai, E.W.T., Moon, K.K., Riggins, F.J., Yi, C.Y.: RFID research: an academic literature review (1995–2005) and future research directions. Int. J. Prod. Econ. **112**(2), 510–520 (2008)
10. Shuhua, M., Yanzhu, H.: Research on the order picking optimization problem of the automated warehouse. In: Proceedings of the IEEE Chinese Control and Decision Conference, pp. 990–993 (2009)
11. Bartholdi III, J.J., Subramanian, S.: Pick-path optimization. http://www2.isye.gatech.edu/jjb/wh/apps/pickpath/pickpath.html (Accessed 25 January 2013)
12. Roodbergen, K.J., De Koster, R.: Routing methods for warehouses with multiple cross aisles. Int. J. Prod. Res. **39**(9), 32–43 (2001)
13. Vaughan, T.S.: The effect of warehouse cross aisles on order picking efficiency. Int. J. Prod. Res. **37**(4), 881–897 (1999)
14. Ratliff, H.D., Rosenthal, A.S.: Order-picking in a rectangular warehouse: a solvable case of the traveling salesman problem. Oper. Res. **31**(3), 507–521 (1983)

15. Tang, H.Y., Li, M.J.: An improved ant colony algorithm for order picking optimization problem in automated warehouse. Fuzzy Inf. Eng. **2**, 1537–1547 (2009)
16. Li, S.Y., Chen, Y.Q., Li, Y.: Ant Colony Algorithms with Applications. Harbin Institute of Technology Press, Harbin (2004)
17. Zhang, G., Lai, K.: Combining path relinking and genetic algorithms for the multiple-level warehouse layout problem. Eur. J. Oper. Res. **169**(2), 413–425 (2006)
18. Huang, X.F., Yun-xia, L.I.U.: Optimization of operational route in AS/RS based on particle swarm algorithm. J. Southwest Jiaotong Univ. (Engl. Ed.) 16(1), Article ID: 1005–2429(2008) 01–0092-03 (2008)
19. Cao, H.Z., Yu, X.C.: Parthenon-genetic simulated annealing algorithm and its application in combinatorial optimization problems. J. Beijing Univ. Posts Telecommun. **3**, 38–41 (2008)
20. Chang-you, L., Yun-wei, S., Jia-heng, L.: Intelligent optimizing methods for vehicles scheduling problems in automated warehouse. In: Proceedings of IEEE Control Conference, Hong Kong, pp. 52–56 (2002)
21. Heskett, J.L.: Cube-per-order index- a key to warehouse stock location. Transp. Distrib. Manag. **3**, 27–31 (1963)
22. Kallina, C., Lynn, J.: Application of the cube-per-order index rule for stock location in a distribution warehouse. Interfaces **7**(1), 37–46 (1976)
23. Roodbergen, K.J., Vis, I.F.: A model for warehouse layout. IIE Trans. **38**(10), 799–811 (2006)
24. Malmborg, C., Krishnakumar, B., Simons, G.: A mathematical overview of warehousing systems with single/dual order-picking cycles. Appl. Math. Model. **12**(1), 2–8 (1988)
25. Hou, Y, Li, L. Wang, L.: Picking Routes optimization of automated warehouse based on parthenon genetic ant colony algorithm, International Conference on Convergence Information Technology. Lecture Notes in Information Technology, vol 19, pp. 198–204 (2012)
26. Hooke, R., Jeeves, T.A.: "Direct search" solution of numerical and statistical problems. J. ACM **8**(2), 212–229 (1961)
27. Chen, L., Langevin, A., Riopel, D.: A tabu search algorithm for the relocation problem in a warehousing System. Int. J. Prod. Econ. **129**, 147–156 (2011)
28. Fumi, A., Scarabotti, L., Schiraldi, M.M.: The effect of slot-code optimization in warehouse order picking. Int. J. Eng. Bus. Manag. **5**(20), 1–10 (2013)
29. Fumi, A., Scarabotti, L., Schiraldi, M.M.: Minimizing warehouse space with dedicated storage policy. Int. J. Eng. Bus. Manag. **5**(21), 1–10 (2013)
30. Battista, C., Fumi, A., Laura, L., Schiraldi, M.M.: Multiproduct slot allocation heuristic to minimize storage space. Int. J. Retail Distrib. Manag. **42**(3), 172–186 (2014)
31. Jensen, T.R., Toft, B.: Graph Coloring Problems. John Wiley & Sons, New York (2011)
32. Marx, D.: Graph colouring problems and their applications in scheduling. Electr. Eng. **48**(1–2), 11–16 (2004)
33. Ho, W., Emrouznejad, A.: Multi-criteria logistics distribution network design using SAS/OR. Expert Syst. Appl. **36**, 7288–7298 (2009)
34. Saaty, T.L.: Decision making with the analytic hierarchy process. Int. J. Serv. Sci. **1**(1), 83–98 (2008)
35. Tamiz, M., Jones, D., Romero, C.: Goal programming for decision making: an overview of the current state-of-the-art. Eur. J. Oper. Res. **111**(3), 569–581 (1998)

36. He, T., Ho, W., Ka Man, C.L., Xu, X.: A fuzzy AHP based integer linear programming model for the multi-criteria transshipment problem. Int. J. Logistics Manag. **23**(1), 159–179 (2012)
37. Kahraman, C., Cebeci, U., Ulukan, Z.: Multi-criteria supplier selection using fuzzy AHP. Logistics Inf. Manag. **16**(6), 382–394 (2003)
38. Tinelli, L.M., Vivaldini, K.C.T., Becker, M.: Intelligent warehouse product position optimization by applying a multi-criteria tool. In: Neto, P., Moreira, A.P. (eds.) WRSM 2013. CCIS, vol. 371, pp. 137–145. Springer, Heidelberg (2013)
39. Ramírez, D.G., Ramírez, L.E., Jiménez, M.A., Castro, L.J., Maldonado, E.: The design of a real-time warehouse management system that integrates simulation and optimization models with RFID technology. Int. J. Comput. Sci. **2**(4), 18–37 (2013)

Reconstruction of the Surface Heat Flux for a Quasi-linear System of the Hyperbolic Type Heat-Conduction Equations

Valentin Borukhov and Olga Kostyukova[✉]

Institute of Mathematics, National Academy of Sciences of Belarus,
Surganov Str. 11, 220072 Minsk, Belarus
{borukhov,kostykova}@im.bas-net.by

Abstract. The problem of the identification of the surface heat flux for a quasi-linear system of the hyperbolic type heat-conduction equations is studied. An approach is proposed based on the stage-by-stage suboptimal optimization of the cost functional and input data filtering using the HuberTikhonov functional. Results are presented for the numerical modeling of the identification problem in conditions of both standard noisy data and noise emissions.

Keywords: Inverse problem · Heat flux · The hyperbolic type system · HuberTikhonov functional · Suboptimal optimization

1 Introduction

To control processes of heat transfer, mass transfer, etc., different methods of mathematical modeling, the theory of optimal control and the theory of inverse problems (IP) of mathematical physics are widely applied [1–4].

In particular, in scientific literature the classical IP theory of heat conductivity consisting in reconstruction of the time-varying heat fluxes is frequently used (see [1–8] and the references therein).

As a rule, the mathematical model of transfer processes is based on the parabolic type heat conductivity equation. At the same time, a number of fast proceeding and intensive transfer processes can be described only within the theory of the hyperbolic type equations and systems of the equations [9–16].

In this paper, we consider IP of reconstruction of the time-varying surface heat flux for quasi-linear system of hyperbolic type differential equations [9,10]. The Dirichlet boundary conditions, initial conditions and time-varying temperature at a given interior point are used as the additional data for reconstruction.

The system under consideration describes transfer processes in the nonlinear mediums and takes into account both the heat flux relaxation time $\tau \geq 0$ and the convective component of the heat transfer [9,10]. When $\tau = 0$, this system of equations can be reduced to the parabolic type heat conductivity equation. Notice that generally, the system is not reducible to one equation [10].

© Springer International Publishing Switzerland 2015
A. Plakhov et al. (Eds.): EmC-ONS 2014, CCIS 499, pp. 49–67, 2015.
DOI: 10.1007/978-3-319-20352-2_4

As it is known, the heat fluxes reconstruction IP belongs to the class of ill-posed problems. Currently, there are several approaches to solving such problems [1–4,20,21]. However, there is no universal method among them that is caused by both difficulties of solving the ill-posed problems and the requirements of an effective realization and high speed of numerical procedures.

In this paper, we develop a method of stage-by-stage suboptimal optimization (SSO) (see [17–19]) combined with a method of the robust estimation on the base of the Huber loss function [22–25].

We consider the filtering procedure as an optimal control problem for the simplest differential equation of the first order [19]. The cost functional for the optimal control problem is the sum of the Huber functional for residuals and the Tikhonov functional for control. Under a suitable choice of settings of the filter, this procedure allows to smooth out and filter out both separate gross measurement errors ("wild" data points, outliers) and standard random errors. It should be mentioned here that the "wild" values of the measurements are quite common in the high-temperature experiences in industry, in the thermal protection methods, in the rocket engines testing, etc.

Notice that the SSO approach develops ideas of the sequential estimation [1,26] that allows to realize data processing in real time. This is important, for example, for the problems of the thermal processes' control.

2   Problem Statement

Let us consider the following initial boundary value problem for the quasi-linear system:

$$\rho(T)C(T)\frac{DT}{Dt} = -\frac{\partial q}{\partial x},$$

$$\tau\frac{Dq}{Dt} = -q - \lambda(T)\frac{\partial T}{\partial x}, \tag{1}$$

$$T = T(x,t), \quad q = q(x,t), \quad x_0 \le x \le x_*, \quad t_0 \le t \le t_*,$$

with initial and boundary conditions

$$T(x,t_0) = T_{in}(x), \quad q(x,t_0) = q_{in}(x), \quad x_0 \le x \le x_*, \tag{2}$$

$$T(x_*,t) = T_*(t), \quad t_0 \le t \le t_*, \tag{3}$$

$$q(x_0,t) = q_0(t), \quad t_0 \le t \le t_*. \tag{4}$$

Here $\dfrac{D}{Dt} = \dfrac{\partial}{\partial t} + \nu(x)\dfrac{\partial}{\partial x}$ is the material derivative, $\nu(x)\dfrac{\partial}{\partial x}$ is convection term, $\nu(x)$ is velocity, $T = T(x,t)$, $x \in [x_0, x_*] \subset \mathbb{R}$, $t \in [t_0, t_*] \subset \mathbb{R}$, is a temperature distribution, $q = q(x,t)$, $x \in [x_0, x_*] \subset \mathbb{R}$, $t \in [t_0, t_*] \subset \mathbb{R}$, is a heat flux, $\lambda(T)$ is the heat conduction coefficient, $\rho(T)$ is the material density, $C(T)$ is the specific heat of the material, τ is a parameter describing the heat flux relaxation time, $T_{in}(x), q_{in}(x), x \in [x_0, x_*]$ are given initial conditions, $T_*(t), t \in [t_0, t_*]$, is a prescribed temperature at the point x_*, $q_0(t), t \in [t_0, t_*]$, is a prescribed heat flux at the point x_0. All the functions $\lambda(T)$, $C(T)$, $\rho(T)$, $T \in \mathbb{R}$, $T_*(t), q_0(t)$, $t \in [t_0, t_*]$, and $T_{\text{in}}(x), q_{\text{in}}(x)$, $\nu(x)$, $x \in [x_0, x_*]$, are supposed to be sufficiently smooth.

In the problem under consideration, there is a convective component of the heat transfer. The presence of this component leads to instability of numerical methods for solving the direct and inverse problems. In particular, the standard **pdepe** -program of the computational MATLAB package doesn't allow to obtain the numerical solution of the direct problem with large values of parameter ν.

Notice that in the case $\nu(x) \equiv 0$, problem (1) can be reduced to a nonlinear heat conduction equation of the hyperbolic type

$$\tau \frac{\partial}{\partial t}\left(C(T)\rho(T)\frac{\partial T}{\partial t}\right) + \rho(T)C(T)\frac{\partial T}{\partial t} = \frac{\partial}{\partial x}\left(\lambda(T)\frac{\partial T}{\partial x}\right)$$

with initial and boundary conditions that can be determined on the base of (2)–(4).

In the paper, for the case $\nu(x) \not\equiv 0$, we consider the problem of reconstruction of the heat flux $q_0(t) := q(x_0, t)$, $t \in [t_0, t_*]$, at the point $x = x_0$ on the base of the known measurements

$$y(t) = T(x_1^*, t) + v(t), \ t \in [t_0, t_*], \tag{5}$$

of the temperature field at a given point x_1^*, $x_0 \leq x_1^* \leq x_*$. Here function $v(t)$ describes a measurement error.

Following [1,2], to model function $v(\cdot)$ we will apply the statistical description. Taking into account the discrete representation of the temperature measurements, this description takes the form

$$y(t_i) = T(x_1^*, t_i) + w(t_i)\sigma, i = 0, 1, ..., M; \tag{6}$$

where

$$t_i = t_0 + i\Delta t, \quad \Delta t = (t_* - t_0)/M, \tag{7}$$

σ is the standard deviation of the measurement errors, $w(t_i)$ is a realization of the random variable w with the normal distribution. Here we consider that the noisy measurements were carried out for the time stepsize $\Delta t > 0$ (see (6)).

In the nonstandard case, we assume that the probability density function of the measurement error has the form [22]

$$f(w) = \frac{1 - \epsilon}{\sqrt{2\pi}} \exp\left(-\frac{w^2}{2}\right) + \epsilon g(w), \tag{8}$$

where ϵ is a weight parameter and $g(w)$ is an unknown function that perturbs the Gaussian density function. Note that the error probability density function of form (8) can be regarded as a model of large errors [22] in the measurement data. Having assumed in (8) that $\epsilon = 0$, one gets the standard situation.

Let the thermophysical parameters $\rho(T)$, $\lambda(T)$, $C(T)$, τ, $\nu(x)$, the temperatures $T_{\text{in}}(x)$, $x \in [x_0, x_*]$, $T_*(t)$, $t \in [t_0, t_*]$, the heat flux $q_{\text{in}}(x)$, $x \in [x_0, x_*]$, the weight parameter ϵ, and the values of the error deviation σ be known. Then, the general problem of identifying the heat flux at the point $x = x_0$ consists in reconstruction of the function $q_0(t) := q(x_0, t)$, $t \in [t_0, t_*]$, from the data (6), system (1) and conditions (2) and (3).

One of the distinguishing characteristics of the proposed approach is that, before solving the posed problem of reconstructing the heat flux from the given inaccurate measurements $y(t)$, $t \in [t_0, t_*]$, a filtering procedure is employed. This procedure yields estimates $y^*(t)$ of the data (5), and the heat flux $q(x_0, t)$, $t \in [t_0, t_*]$, is reconstructed on the basis of these estimates. The filtering procedure is described in Sect. 3. The second specific feature of the proposed approach is that the problem of reconstructing the heat flux $q(x_0, t)$, $t \in [t_0, t_*]$, is solved by the SSO method [18,19]. The essence and advantages of the method are described in Sect. 4. It should be noted that the ideas of this method are also used in the prefiltering procedure.

3 Preliminary Filtering Procedure Using the Huber Function and Tikhonov Regularization

Let a given function $y(t)$, $t \in [t_0, t_*]$, be representable in the form

$$y(t) = y^0(t) + v(t), \ t \in [t_0, t_*], \tag{9}$$

where $y^0(t)$, $t \in T$, is some unknown smooth function, $v(t)$, $t \in [t_0, t_*]$, is a function of unknown disturbances (a noise). It is required, to find a continuous smooth function $y^*(t)$, $t \in [t_0, t_*]$, which approximates the function $y^0(t)$, $t \in [t_0, t_*]$, on the basis of the given noisy function (9).

To solve this problem, taking into account the smoothness of the reconstructed function $y^0(t)$, $t \in [t_0, t_*]$, we formulate the simplest optimal control problem

$$\int_{t_0}^{t_*} f(x(t) - y(t))dt + R(u(\cdot)) \to \min_{z, u(\cdot)} \tag{10}$$

$$\text{s.t. } \dot{x}(t) = u(t), \ x(0) = z.$$

Here $f(z)$ is some function that characterizes the deviation of $|z|$ from zero, $R(u(\cdot))$ is a regularizing term. The particular choice of the functions $f(z)$ and $R(u(\cdot))$ depends on the a priori information about the restored function $y^0(t)$, $t \in [t_0, t_*]$, and the nature of the unknown noise $v(t)$, $t \in [t_0, t_*]$. Most often the l_1- and l_2-norms are used as $f(z)$ and the functional $\beta \int_{t_0}^{t_*} u^2(t)dt$ is used as the regularizing term $R(u(\cdot))$ [21].

In this paper (see also [19]), we propose to use the Huber loss function as the deviation function $f(z)$ [22]. This function is a combination of l_1- and l_2- norms.

It is known that the Huber function is robust in the sense that it can reduce the influence of "wild" data points (outliers). The Huber function has the form

$$f_\gamma(z) = \begin{cases} z^2/2 & \text{if } |z| \leq \gamma, \\ \gamma|z| - \gamma^2/2 & \text{if } |z| > \gamma. \end{cases}$$

It is quadratic if the module of the deviation z is smaller than a given constant $\gamma > 0$ and has an absolute value term if the module of the deviation is greater than γ.

As it is known (see [22]), the tuning parameter γ is related with the perturbing parameter ϵ by means of the implicit equation

$$\frac{1}{1 - \varepsilon} = \frac{1}{\gamma}\sqrt{\frac{2}{\pi}} \exp\left[-\frac{\gamma^2}{2}\right] + \text{erf}\left(\frac{\gamma}{\sqrt{2}}\right).$$

The Huber function is more robust than the l_2-function in the sense that it is less sensitive to the outliers in the measurement data.

The functional

$$R_{\alpha_*,\beta_*}(u(\cdot)) := \beta_* \int_{t_0}^{t_*} u^2(t)dt + \eta_* \int_{t_0}^{t_*} \dot{u}^2(t)dt$$

can be used as a regularizing term. Here $\alpha_* \geq 0, \beta_* \geq 0$ are the weight coefficients.

Thus, in order to generate the estimate $y^*(t), t \in [t_0, t_*]$, of the unknown function $y^0(t), t \in [t_0, t_*]$, on the basis of the noisy function $y^*(t), t \in [t_0, t_*]$, we solve the following optimal control problem:

$$\int_{t_0}^{t_*} f_\gamma(x(t) - y(t))dt + \beta_* \int_{t_0}^{t_*} u^2(t)dt + \eta_* \int_{t_0}^{t_*} \dot{u}^2(t)dt, \rightarrow \min_{z,u(\cdot)}, \qquad (11)$$

$$\dot{x}(t) = u(t), \quad x(0) = z.$$

Let $z^0, u^0(t), t \in [t_0, t_*]$, be an optimal solution of problem (11). Then the function

$$y^*(t) = z^0 + \int_{t_0}^{t} u^0(\tau)d\tau, \ t \in [t_0, t_*],$$

is considered as an approximation of the unknown function $y^0(t), t \in [t_0, t_*]$.

To solve problem (11), we apply a method of a stage-by-stage optimization. The idea of the method is identical to that of the method described in [17–19]. The essence of the method consists in the reduction of the filtering procedure over the entire time interval $t \in [t_0, t_*]$ to the consecutive solution of p filtering problems on small time intervals $t \in [\tau_j, \tau_{j+1}], j = 0, 1, ..., p - 1$, where $\tau_j = t_0 + jL\Delta t$, is a time instant from the set $\{t_i, i = 0, 1, ..., M\} \subset [t_0, t_*], L > 0, p > 0$ are integers. Here $L\Delta t$ is the length of the stage defined by the parameter L and a time stepsize $\Delta t > 0$. The time stepsize $\Delta t > 0$ is a step with which the noisy measurements were carried out (see (6)).

Given j $(j = 0, ..., p - 1)$, to find an approximation $y^*(t)$ of the unknown function $y^0(t)$ at the interval $t \in [\tau_j, \tau_{j+1}]$, on the base of the noisy data $y(t_i)$, $t_i \in [\tau_j, \tau_{j+1}]$, $i = jL, ..., (j + 1)L$, the following optimal control problem is solved.

Problem F_j: It is required to find a control $u(t), t \in [\tau_j, \tau_{j+1}]$, that solves the problem

$$\sum_{i=jL}^{(j+1)L} f_\gamma(x(t_i) - y(t_i)) + \beta_* \int_{\tau_j}^{\tau_{j+1}} u^2(t)dt + \eta_* \int_{\tau_j}^{\tau_{j+1}} \dot{u}^2(t)dt +$$

$$\bar{\eta}_*(z - y^*(\tau_j - 0))^2 \to \min_{z,u(\cdot)}$$

$$\text{s.t. } \dot{x}(t) = u(t), \quad x(\tau_j) = z.$$

Here, in the cost functional, we add one additional term $(z - y^*(\tau_j - 0))^2$, where $y^*(\tau_j - 0) := \lim_{t \to \tau_j, t < \tau_j} y^*(t)$, with the weighting coefficient $\bar{\eta}_* \geq 0$. This term is responsible for "continuous matching" of the functions $y^*(t), t \in [\tau_{j-1}, \tau_j]$, and $y^*(t), t \in [\tau_j, \tau_{j+1}]$, being the optimal trajectories in problems F_{j-1} and F_j respectively, at the boundary point τ_j for two adjacent subintervals. For $j = 0$, we set $y^*(\tau_j - 0) = y(t_0)$.

Let $u^0(t)$, $t \in [\tau_j, \tau_{j+1}]$, z^{0j} be an optimal solution in problem F_j. Then we set

$$y^*(t) = z^{0j} + \int_{\tau_j}^{t} u^0(\tau)d\tau, \quad t \in [\tau_j, \tau_{j+1}],$$

and consider $y^*(t), t \in [\tau_j, \tau_{j+1}]$, as a result of filtering the noisy data $y(t)$ at the j-th state.

For the numerical purposes, we consider the problem F_j in a class of the piecewise constant controls

$$u(t) = u_i, \, t \in [t_i, t_{i+1}], \quad i = jL, jL + 1, ..., (j + 1)L - 1, \tag{12}$$

$$t_i = t_0 + i\Delta t, \quad \Delta t = (t_* - t_0)/M, \quad M = (p - 1)L.$$

In that case, the problem F_j is equivalent to a quadratic programming problem that can be easy solved by a standard quadratic programming solver.

4 Heat Flux Reconstruction by the SSO Method

The problem of reconstructing the heat flux $q(x_0, t), t \in [t_0, t_*]$, is formulated as an optimal control problem in which the role of the sought control is played by the reconstructed heat flux and the purpose of optimization is to minimize the functional of the squared deviation between the calculated states $T(x_1^*, t_i)$ of system (1) and the data $y^*(t_i)$, $i = 1, ..., M$. The optimal control problem is solved by the SSO method.

As it was already mentioned above, the idea of the method consists in reducing the single problem of reconstructing the heat flux $q(x_0, t)$ over the entire time interval $t \in [t_0, t_*]$ to the succession of p problems of reconstructing this flux on the small intervals $t \in [\tau_j, \tau_{j+1}]$, $j = 0, 1, ..., p-1$, where $\tau_j = t_0 + jL\Delta t$ and integers $L > 0$, $p > 0$, $M > 0$ are the same as in Sect. 3.

Let us describe the main steps of the method.

Given j ($j = 0, ..., p-1$), let us suppose that the heat flux $q_0(t) := q(x_0, t)$ has been restored for $t \in [t_0, \tau_j]$ and the noisy data $y(t_i)$, $t_i \in [t_0, \tau_j]$, $i = 1, ..., jL$, have been filtered. Hence we know the function $q_0^*(t)$, $t \in [t_0, \tau_j]$, approximating the heat flux $q_0(t)$, $t \in [t_0, \tau_j]$, and the function $y^*(t)$, $t \in [t_0, \tau_j]$. This gives us opportunity to find the solution $T^*(x, t)$, $q^*(x, t)$, $x \in [x_0, x^*]$, $t \in [t_0, \tau_j]$, of system (1)–(4) with $q_0(t)$, $t \in [t_0, \tau_j]$, replaced by $q_0^*(t)$, $t \in [t_0, \tau_j]$.

Using the known number $y^*(\tau_j - 0)$ and the functions $T^*(x, \tau_j)$, $q^*(x, \tau_j)$, $x \in [x_0, x^*]$, we consider the problem of reconstructing the heat flux $q(x_0, t)$ on the interval $t \in [\tau_j, \tau_{j+1}]$.

For this purpose, first of all, we apply the filtering procedure described in Sect. 3 to the known number $y^*(\tau_j - 0)$ and the known noisy data $y(t_i)$, $t_i \in [\tau_j, \tau_{j+1}]$, $i = jL+1, ..., (j+1)L$. To do this, we have to solve problem F_j. As a result, we obtain the estimates $y^*(t)$, $t \in [t_j, t_{j+1}]$, of the noisy measurements.

In order to reconstruct the heat flux $q(x_0, t)$, $t \in [\tau_j, \tau_{j+1}]$, from the new data $y^*(t)$, $t \in [\tau_j, \tau_{j+1}]$, and the known function $T^*(x, \tau_j - 0)$, $q^*(x, \tau_j - 0)$, $x \in [x_0, x_*]$, the following optimal control problem is solved.

Problem P_j: find a control $U(t)$, $t \in [\tau_j, \tau_{j+1}]$, which minimizes the cost functional

$$\sum_{i=jL+1}^{(j+1)L} (T(x_1^*, t_i) - y^*(t_i))^2 + \eta_j \int_{\tau_j}^{\tau_{j+1}} \left(\frac{dU(t)}{dt}\right)^2 dt +$$

$$\gamma_j (U^*(\tau_j - 0) - U(\tau_j + 0))^2 \to \min \qquad (13)$$

on the trajectories $T(x, t)$, $q(x, t)$, $x \in [x_0, x_*]$, $t \in [\tau_j, \tau_{j+1}]$, of the system

$$\rho(T)C(T)\frac{DT}{Dt} = -\frac{\partial q}{\partial x},$$

$$\tau\frac{Dq}{Dt} = -q - \lambda(T)\frac{\partial T}{\partial x}, \qquad (14)$$

$$T = T(x, t), \quad q = q(x, t), \quad x_0 \le x \le x_*, \quad \tau_j \le t \le \tau_{j+1},$$

$$T(x, \tau_j) = T^*(x, \tau_j - 0), \quad q(x, \tau_j) = q^*(x, \tau_j - 0), \quad x_0 \le x \le x_*, \qquad (15)$$

$$T(x_*, t) = T_*(t), \quad q(x_0, t) = U(t), \quad \tau_j \le t \le \tau_{j+1}. \qquad (16)$$

Here, in the cost functional (13), the second term is a Tikhonov type regulator with a weighting coefficient $\eta_j > 0$. The third term $\gamma_j(U^*(\tau_j - 0) - U(\tau_j + 0))^2$ is the penalty term (with a weighting coefficient $\gamma_j > 0$) which is responsible for matching the boundary values $U^*(\tau_j - 0)$ and $U^*(\tau_j + 0)$ of the controls obtained at the neighboring $(j-1)$-th and j-th stages; $U^*(t), T^*(x, t), q^*(x, t), t \in [\tau_{j-1}, \tau_j]$, $x \in [x_0, x_*]$, are the optimal control, the corresponding temperature and the heat flux obtained on the previous $(j-1)$-th stage. For $j = 0$ we consider $\gamma_0 = 0$ and $T^*(x, \tau_0 - 0) = T_{in}(x), q^*(x, \tau_0 - 0) = q_{in}(x)$, $x \in [x_0, x_*]$.

Let $U^*(t)$ and $T^*(x,t)$, $q^*(x,t)$, $t \in [\tau_j, \tau_{j+1}]$, $x \in [x_0, x_*]$, be an optimal control and the corresponding trajectory in the problem P_j. Then we set

$$q_0^*(t) = q^*(x_0, t), \quad t \in [\tau_j, \tau_{j+1}], \tag{17}$$

and consider $q_0^*(t)$, $t \in [\tau_j, \tau_{j+1}]$, as an approximation of the recoverable heat flux $q(x_0, t)$ at the j-th state.

To solve problem P_j numerically, the nonlinear system of partial differential equations (14)–(16) is approximated by a system of ordinary differential equations. For this purpose, let us partition the interval $[x_0, x_*]$ into N parts by the points

$$x^i = x_0 + i\Delta x, \ i = 0, 1, ..., N, \ \ \Delta x = (x_* - x_0)/N, \tag{18}$$

$$x_1^* = x_{i_*} = x_0 + i_* \Delta x.$$

Here and in what follows, without loss of generality, we consider that the point x_1^*, at which the temperature's measurements were performed, is a node of the grid (18).

Denote

$$z_1(t) = T_1(t) = T(x_1, t), \ z_{2i} = T_{i+1}(t) = T(x_{i+1}, t), \ i = \overline{1, N-2};$$
$$z_{2i-1}(t) = q_i(t) = q(x_i, t), i = \overline{2, N-1}; \ z_{2N-2}(t) = q_N(t) = q(x_N, t),$$

and consider a vector-function $Z(t) = (z_1(t), ..., z_{2N-2}(t))$, $t \in [\tau_j, \tau_{j+1}]$. Then the cost functional (13) takes the form

$$\sum_{i=jL+1}^{(j+1)L} (z_k(t_i) - y^*(t_i))^2 + \eta_j \int_{\tau_j}^{\tau_{j+1}} \left(\frac{dU(t)}{dt}\right)^2 dt +$$

$$\gamma_j(U^*(\tau_j - 0) - U(\tau_j + 0))^2 \to \min, \tag{19}$$

where $k = 2(i_* - 1)$ if $1 < i_* \le N - 1$, and $k = 1$ if $i_* = 1$.

If N is a rather large number, the system (14)–(16) can be approximated by the following nonlinear system of ordinary differential equations:

$$\frac{dZ(t)}{dt} = \bar{F}(Z(t), \ T_*(t), \ U(t)), \ \ Z(t_0) = Z_0 = (z_1^0, ..., z_{2N-2}^0), \tag{20}$$

with

$$z_1^0 = T^*(x_1, \tau_j), \ z_{2i}^0 = T^*(x_{i+1}, \tau_j), \ i = 1, ..., N - 2;$$
$$z_{2i-1}^0 = q^*(x_i, \tau_j), \ i = 2, ..., N - 1; \ z_{2N-2}^0 = q^*(x_N, \tau_j); \tag{21}$$

$$\bar{F}(Z, \ T_*(t), \ U) = F(Z, U, t) = (F_i(Z, U, t), i = 1, ..., 2(N-1)),$$

$$F_1(Z, U, t) = -\nu_1 \frac{z_2 - z_1}{\Delta x} - \frac{z_3 - U}{\Delta x C(z_1)\rho(z_1)};$$

$$F_{2i}(Z, U, t) = -\nu_{i+1} \frac{z_{2(i+1)} - z_{2i}}{\Delta x} - \frac{z_{2i+3} - z_{2i+1}}{\Delta x C(z_{2i})\rho(z_{2i})}; \ i = 1, ..., N - 3;$$

$$F_{2(N-2)}(Z, U, t) = -\nu_{N-1} \frac{T_*(t) - z_{2(N-2)}}{\Delta x} - \frac{z_{2N-2} - z_{2N-3}}{\Delta x C(z_{2(N-2)}) \rho(z_{2(N-2)})};$$

$$F_3(Z, U, t) = -\nu_2 \frac{z_3 - u}{\Delta x} - \left(z_3 + \lambda(z_2) \frac{z_2 - z_1}{\Delta x} \right) \frac{1}{\tau};$$

$$F_{2i-1}(Z, U, t) = -\nu_i \frac{z_{2i-1} - z_{2i-3}}{\Delta x} - \left(z_{2i-1} + \lambda(z_{2(i-1)}) \frac{z_{2(i-1)} - z_{2(i-2)}}{\Delta x} \right) \frac{1}{\tau},$$

$$i = 3, ..., N - 1;$$

$$F_{2N-2}(Z, U, t) = -\nu_N \frac{z_{2N-2} - z_{2N-3}}{\Delta x} - \left(z_{2N-2} + \lambda(T^*(t)) \frac{T_*(t) - z_{2(N-2)}}{\Delta x} \right) \frac{1}{\tau},$$

where $\nu_i = \nu(x_{i-1})$, $i = 1, 2, ..., N$.

Problem (19)–(21) is an optimal control problem for the nonlinear dynamic system (20) with a $(2N - 2)$–dimensional state vector $Z(t) = (z_1(t), ..., z_{2N-2}(t))$, $t \in [\tau_j, \tau_{j+1}]$, and a scalar control $U(t)$, $t \in [\tau_j, \tau_{j+1}]$. This problem has a number of specific features, which make it impossible to use the standard computing packages meant for solving "standard" optimal control problems. Therefore we will make some simplifications.

First of all, taking into account that the interval $[\tau_j, \tau_{j+1}]$ is small, we linearize system (20) on the interval $[\tau_j, \tau_{j+1}]$, having replaced $\lambda(z_i(t))$, $C(z_i(t))$, $\rho(z_i(t))$, $i = 1, ..., 2N - 2$, by $\lambda(z_i^0)$, $C(z_i^0)$, $\rho(z_i^0)$, $i = 1, ..., 2N - 2$. Remind that here the vector $Z(t_0) = Z_0 = (z_1^0, ..., z_{2N-2}^0)$ is defined according to (21), i.e. it is considered to be known at the moment τ_j.

Besides, we will solve this problem in a class of piecewise constant admissible controls

$$U(t) = U_i = const, \ t \in [t_i, t_{i+1}], i = jL, ..., (j + 1)L - 1. \qquad (22)$$

Denote the linearized problem (19)–(21) with additional conditions (22) by P_j^{linear}. The problem P_j^{linear} can be easily reduced (see, for example, [17]) to a quadratic programming problem and solved by standard methods.

Let $U(t) = U^*(t), t \in [\tau_j, \tau_{j+1}]$, be an optimal control in linear quadratic problem P_j^{linear}. Using this control we integrate the system of partial differential equations (14)–(16) with $U(t) = U^*(t)$, $t \in\in [\tau_j, \tau_{j+1}]$. As a result we obtain the trajectory $T^*(x, t)$, $q^*(x, t)$, $x \in [x_0, x_*]$, $t \in [\tau_j, \tau_{j+1}]$. Knowing the trajectory, we set $q_0^*(t) = q^*(x_0, t)$, $t \in [\tau_j, \tau_{j+1}]$, and consider this function as an approximation of the heat flux $q(x_0, t)$, $t \in [\tau_j, \tau_{j+1}]$, obtained at the jth stage.

Using new vector and functions

$$y^*(\tau_{j+1} - 0), \ T^*(x, \tau_{j+1} - 0), \ q^*(x, \tau_{j+1} - 0), \ x \in [x_0, x_*], \qquad (23)$$

we go to the next $(j + 1)$th stage whose aim is to reconstruct the heat flux $q(x_0, t), t \in [\tau_{j+1}, \tau_{j+2}]$, on the base of data (23) and the noisy temperature measurements $y(t_i)$ at the time instants $t_i \in [\tau_{j+1}, \tau_{j+2}]$, $i = (j + 1)L + 1, ...,$ $(j + 2)L$.

As it was shown in [17,18], for the optimal control problem P_j^{linear}, the value of index i_* (see relation (18)) is of great significance since it defines the index k of the cost functional of the problem. Notice that $i_* = N(x_1^* - x_0)/(x_* - x_0)$ is uniquely defined by the given values x_0, x_1^*, x_* and a chosen parameter N.

We recall [27] that the index of the cost functional

$$\int_a^b f(Z(t), U(t), t)dt$$

is the smallest integer number k such that $\frac{\partial}{\partial U} \frac{d^k}{dt^k} f(Z(t), U(t), t) \neq 0$. Here the derivatives $\frac{d^k Z(t)}{dt^k}$ are calculated taking into account a specified system of differential equations. In the case under consideration this system coincides with the linearized system (20).

The index k characterizes the degree of the direct influence of a control $U(t)$, $t \in [\tau_j, \tau_{j+1}]$, on the cost functional. The higher the value of the index k, the weaker the influence of $U(t)$, $t \in [\tau_j, \tau_{j+1}]$, on the cost functional (or rather on its first term that is responsible for the restoration's quality) and the more "irregular" the restoration problem becomes. For the problem P_j^{linear}, the indices i_* and k are related as follows:

$$k = 2(i_* - 1) \text{ if } 1 < i_* \text{ and } k = 1, \text{if } i_* = 1.$$

Notice that for the problems considered in papers [17,18] the values of the indices i_* and k coincide: $k = i_*$. It illustrates once again that the identification problems considered in this paper are more difficult than the ones studied in [17,18].

The study of the problem P_j^{linear}, for large index i_* values, shows that the values of the control function $U(t) = U^{(j)}(t), t \in [\tau_j, \tau_{j+1}]$, that are situated closer to the end of the interval $[\tau_j, \tau_{j+1}]$, exert the smallest influence on the first term $\sum_{i=jL+1}^{(j+1)L} (z_k(t_i) - y^*(t_i))^2$ of the cost function: the closer control to the end of the interval, the weaker its influence. The choice of these control values is carried out mainly just for the purpose of minimization of the regularizing term $\int_{\tau_j}^{\tau_{j+1}} (dU(t)/dt)^2 dt$ in the cost functional (19). It is clear that these control values will be "regular", but far from the values of the restored function.

To overcome the specified difficulties arising for large values of the index i_*, it is necessary to insert the following changes to the described above algorithm.

Let us select one more integer parameter L_b, $0 \leq L_b \leq L$. The value of L_b specifies the part $[\tau_j, \tau_j + L_b \Delta t]$ (called confidence interval) of the interval $[\tau_j, \tau_{j+1}]$ where the obtained control actions are supposed to be restored correctly. Only this part, $U^*(t), t \in [\tau_j, \tau_j + L_b \Delta t]$, of the obtained control function $U^*(t), t \in [\tau_j, \tau_j + L \Delta t]$, defined on the confidence interval will be used on the subsequent steps of the algorithm.

Taking into account these changes, the algorithm becomes as follows.

Step 0 (Initialization). Set $j = 0$, $\bar{\tau}_0 = t_0$, $T^*(x, \bar{\tau}_0 - 0) = T_{\text{in}}(x)$, $q^*(x, \bar{\tau}_0 - 0) = q_{\text{in}}(x)$, $x \in [x_0, x_*]$, $y^*(\bar{\tau}_0 - 0) = y(t_0)$.

Step 1. Using the known vector $y^*(\bar{\tau}_j - 0)$, apply filtering procedures to the noisy data $y(t)$, $t \in [\bar{\tau}_j, \tau^*_{j+1}]$, where $\tau^*_{j+1} = \min\{\bar{\tau}_j + L\Delta t, \ t_*\}$, and get function $y^*(t), t \in [\bar{\tau}_j, \tau^*_{j+1}]$.

Step 2. Set

$$z^0_1 = T^*(x_1, \bar{\tau}_j), \ \ z^0_{2i} = T^*(x_{i+1}, \bar{\tau}_j), \ \ i = 1, \ldots, N-2;$$
$$z^0_{2i-1} = q^*(x_i, \bar{\tau}_j), \ i = 2, \ldots, N-1 \ z^0_{2N-2} = q^*(x_N, \bar{\tau}_j). \tag{24}$$

Using the filtered data $y^*(t), t \in [\bar{\tau}_j, \tau^*_{j+1}]$, solve the optimal control problem $\mathrm{P}^{\text{linear}}_j$ on the interval $[\bar{\tau}_j, \tau^*_{j+1}]$. Notice that in (19) one should replace τ_j, τ_{j+1} by $\bar{\tau}_j$, τ^*_{j+1} and $\sum^{(j+1)L}_{i=jL+1}$ by $\sum^{m(\tau^*_{j+1})}_{i=m(\bar{\tau}_j)}$ where $m(\tau) \in \{0, 1, ..., M\}$ with $\tau \in \{t_j, j = 0, 1, ..., M\}$ is such integer number that $\tau = t_{m(\tau)}$. Let $U^*(t), t \in [\bar{\tau}_j, \tau^*_{j+1}]$, be the optimal control of the problem.

Step 3. Set $\bar{\tau}_{j+1} := \bar{\tau}_j + L_b\Delta t$. Integrate the nonlinear system of partial differential equations (14)–(16) on the interval $[\bar{\tau}_j, \bar{\tau}_{j+1}] \subset [\bar{\tau}_j, \tau^*_{j+1}]$ replacing τ_j, τ_{j+1} and $U(t), t \in [\tau_j, \tau_{j+1}]$, by $\bar{\tau}_j$, $\bar{\tau}_{j+1}$ and $U^*(t), t \in [\bar{\tau}_j, \bar{\tau}_{j+1}]$. This yields the trajectory $T^*(x, t)$, $q^*(x, t)$, $x \in [x_0, x_*]$, $t \in [\bar{\tau}_j, \bar{\tau}_{j+1}]$.

Step 4. Set $q^*_0(t) = q^*(x_0, t), t \in [\bar{\tau}_j, \bar{\tau}_{j+1}]$.

Step 5. If $\bar{\tau}_{j+1} = t_*$, go to Step 7, otherwise go to Step 6.

Step 6. Set $j := j + 1$ and go to Step 1.

Step 7. The Algorithm stops the work.

The constructed function $q^*_0(t), t \in [t_0, t_*]$, is taken as the restored heat flux $q(x_0, t)$, $t \in [t_0, t_*]$ at the point x_0. The described algorithm is consistent with the approach based on the sequential estimation [1,26].

Thus, in the proposed method, the process of solving a single reconstruction problem for nonlinear system (1)–(4) on the large interval is reduced to the process of solving a succession of p optimal control problems $\mathrm{P}^{\text{linear}}_j$ for linear systems on small intervals $t \in [\tau_j, \tau_{j+1}]$, $j = 0, ..., p - 1$. It should be noted that, for a one-stage reconstruction procedure, i.e., when $p = 1$, only one problem of the optimal control is solved on the entire interval $[t_0, t_*]$. However, the dimensionality of this problem grows up as the discretization steps Δt and Δx decrease, which makes it impossible to solve this problem with high accuracy. In the proposed approach for arbitrarily small values of the steps Δt and Δx, the dimensions of the quadratic programming problems to be solved at each stage may take any prescribed values. For fixed values of dimensionality of these problems, the reduction of the discretization steps Δt and Δx results only in the increase of the number p of stages.

Note also that the small lengths of the intervals $[\tau_j, \tau_{j+1}]$, $j = 0, 1, ..., p - 1$, (determined by the parameters L and M) and the stage-by-stage character of the SSO method enable the user to circumvent effectively the difficulties associated with the nonlinearity of the system. The fact that the algorithm includes the confidence interval determined by the parameter L_b allows one to reduce the difficulties associated with the irregularity of the problem for large values of the index i_*.

Besides, at each stage it is possible to analyze the quality of restoration and to correct the regularization parameters at the next stage on the basis of the results of this analysis.

5   Numerical Modeling

First of all, to evaluate the quality of approximation of the system of partial differential equations (1)–(4) by the system of ordinary differential equations (20), we solve two direct problems.

The first problem consists in constructing the functions

$$T(x,t), \ q(x,t), \ x_0 \leq x \leq x_*, \ t_0 \leq t \leq t_*, \tag{25}$$

that satisfy system (1) and the given initial and boundary conditions (2)–(4) with the following parameter's values

$$C(T) = 1, \ \tau = 2, \ \lambda(T) = 1, \ \rho(T) = 5, \ \nu(x) = 0.1,$$
$$x_0 = 0, \ x_* = 5, \ t_0 = 0, \ t_* = 50, \tag{26}$$

and functions $q_{in}(x)$, $T_{in}(x)$, $x \in [x_0, x_*]$, and $q_0(t)$, $T_*(t)$, $t \in [t_0, t_*]$, that are presented in Fig. 1(a) and (b), respectively. Here the plots of the functions $q_{in}(x)$ and $q_0(t)$ are denoted by the continuous lines and the plots of the functions $T_{in}(x)$ and $T_*(t)$ are denoted by the dotes.

Notice that the initial and boundary conditions (2)–(4) satisfy the following consistency constraints:

$$T_{in}(x_*) = T_*(t_0), \ q_{in}(x_0) = q_0(t_0), \tag{27}$$

$$\rho(T_*(t_0))C(T_*(t_0)) \left(\frac{dT_*(t_0)}{dt} + \nu(x_*)\frac{dT_{in}(x_*)}{dx} \right) = -\frac{dq_{in}(x_*)}{dx} \tag{28}$$

$$\tau \left(\frac{dq_0(t_0)}{dt} + \nu(x_0)\frac{dq_{in}(x_0)}{dx} \right) = -q_{in}(x_0) - \lambda(T_{in}(x_0))\frac{dT_{in}(x_0)}{dx}. \tag{29}$$

To solve the first direct problem, the standard **pdepe**-program of the computing package MATLAB was used.

For the given data, the plots of functions (25) are shown in Fig. 2.

After that, for the same set of parameters (26) and functions $q_{in}(x)$, $T_{in}(x)$, $x \in [x_0, x_*]$, $q_0(t)$, $T_*(t)$, $t \in [t_0, t_*]$, we have solved the initial value problem for system (20) in which the state vector $Z(t) = (z_1(t), ..., z_{2N-2}(t))$ (with $N = 11$) had the form

$$z_1(t) = T_1(t) = T(x_1, t), \ z_{2i} = T_{i+1}(t) = T(x_{i+1}, t), \ i = 1, \ldots, N-2;$$
$$z_{2i-1}(t) = q_i(t) = q(x_i, t), \ i = 2, \ldots, N-1; \ z_{2N-2}(t) = q_N(t) = q(x_N, t).$$

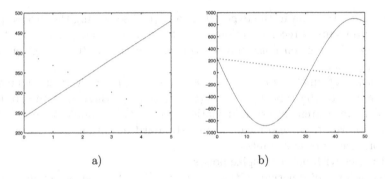

a)   b)

Fig. 1. (a) Functions $q_{in}(x)$, $T_{in}(x)$, $x \in [x_0, x_*]$; (b) functions $q_0(t)$, $T_*(t)$, $t \in [t_0, t_*]$.

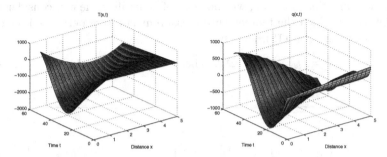

Fig. 2. Functions $T(x, t)$, $q(x, t)$, $t \in [t_0, t_*]$, $x \in [x_0, x_*]$.

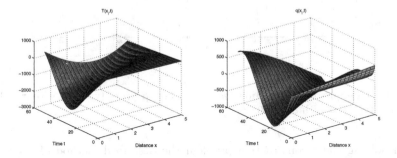

Fig. 3. Functions $T_i(t)$, $q_i(t)$, $t \in [t_0, t_*]$, $i = 1, ..., N$.

The plots of the functions

$$T_i(t) = T(x_i, t), \; q_i(t) = q(x_i, t), \; i = 1, \ldots, N, \; t_0 \le t \le t_*,$$

are presented in Fig. 3.

The carried out numerical calculations showed that for a rather large value of the parameter N, the system of the ordinary differential equations (20) well approximates the initial system of partial differential equations (1)–(4).

The main attention in the experiment was paid to solving the inverse problems, i.e. problems of reconstruction of the function $q_0(t) = q(x_0, t)$, $t \in [t_0, t_*]$, on the base of the given noisy temperature measurements (6) at a given point x_1^*, $x_0 \le x_1^* \le x_*$.

Two types of functions $v(t)$, $t \in [t_0, t_*]$, modeling noises were considered:

(A) noises of the type $v(t) = \sigma w(t)$, $t \in [t_0, t_*]$, where σ is the standard value of the measurement errors deviation, and $w(t)$ is a random variable with a normal distribution, zero mean, a unit standard deviation, and uncorrelated values for various time instants;

(B) noises with outliers (spike noises).

The examples of functions $v(t) = \sigma w(t)$, $t \in [t_0, t_*]$, of types (A) and (B) are presented in Fig. 4.

As it was mentioned before, without loss of generality one can consider that the point x_1^*, at which the temperature measurements were performed, belongs to the nodes of the grid (18), namely,

$$x_1^* = x_{i_*} = x_0 + i_* \Delta x, \text{ where } i_* \in \{1, ..., N-1\}. \tag{30}$$

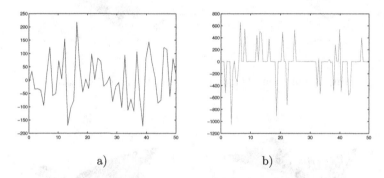

a)   b)

Fig. 4. (a) Examples of the function $\sigma w(t)$, $t \in [t_0, t_*]$: (a) of type (A) with $\sigma = 100$; and (b) of type (B) with $\sigma = 400$

For solving the reconstruction problem the described method of SSO was applied.

For different levels and types of noise σ and different values of the point $x_1^* = x_{i_*}$, the results of reconstruction of the function $q_0(t)$, $t \in [t_0, t_*]$, are presented in Figs. 5, 6, 7 and 8. Here the model function $q_0(t)$, $t \in [t_0, t_*]$, and functions $q_0^*(t)$, $t \in [t_0, t_*]$, obtained as a result of application of the described reconstruction method are shown. The parameter N was chosen to be equal to 11.

In the top parts of Figs. 5, 6, 7 and 8, the functions of the model heat flux $q_0(t) = q(x_0, t)$, $t \in [t_0, t_*]$, (the dashed line) and the reconstructed fluxes $q_0^*(t)$, $t \in [t_0, t_*]$, (the dot line) are represented.

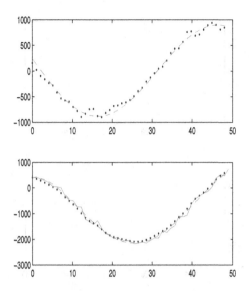

Fig. 5. The results of numerical modeling for reconstruction of heat flux with $i_* = 1$ ($L = 30, \alpha_u = 1.5, \alpha_1 = 2$) and noise of type (A) ($\sigma = 100$)

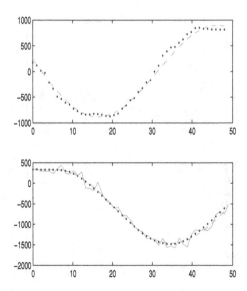

Fig. 6. The results of numerical modeling for reconstruction of heat flux with $i_* = 5$ ($L = 10, \eta_j = 4, \gamma_j = 4$) and noise of type (A) ($\sigma = 100$)

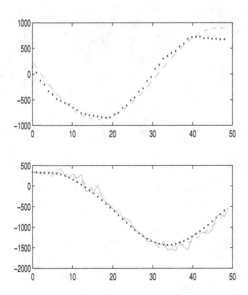

Fig. 7. The results of numerical modeling for reconstruction of heat flux with $i_* = 5$ ($L = 30, \eta_j = 1, \gamma_j = 1$) and noise of type (A) ($\sigma = 100$)

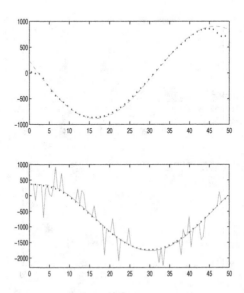

Fig. 8. The results of numerical modeling for reconstruction of heat flux with $i_* = 3$ ($L = 30, \eta_j = 0.25, \gamma_j = 0.25$) and noise of type (B) ($\sigma = 400$)

In the lower parts of these figures, the following functions are presented:

- function of temperature $T(x_1^*, t)$, $t \in [t_0, t_*]$, obtained as a result of solving the direct problem with data (26) and the functions presented on Fig. 1 (this function is denoted by the dashed lines);
- function of temperature obtained as a result of solving the direct problem with the same data in which the model function of $q_0(t)$, $t \in [t_0, t_*]$, is replaced by the function of $q_0^*(t)$, $t \in [t_0, t_*]$, constructed by the offered restoration procedure (this function is denoted by the dot line);
- and also function $y(t)$, $t \in [t_0, t_*]$, (5) (it is denoted by the continuous line).

The performed experiment shows that for systems (1)–(4) the method of stage-by-stage optimization is effective.

The analysis of the obtained results shows that the quality of reconstruction significantly depends on the value of the point x_1^*, in which the temperature's measurements are performed (i.e. on a relation of the parameters i_* and N) and on the noise level σ.

6 Conclusion

In the paper, we have developed the method of the stage-by-stage suboptimal optimization (SSO method) for solving the inverse heat conduction problems. The problem of reconstruction of the time-varying surface heat flux on the base of the temperature measurements in an internal point was studied. We considered the hyperbolic type systems taking into account the heat flux relaxation time, conductive and convective components of the heat distribution processes. Such systems generalize the studied earlier ones and are rather difficult.

The mathematical model of the heat transmission takes into account the medium nonlinearity, the speed of heat convection, and the heat flux relaxation time. For filtering the noisy input data, the method of robust estimation, based on the Tikhonov-Huber cost functional, was applied.

The numerical experiments performed have confirmed the effectiveness of the proposed approach for solving reconstruction problems on the base of both standard noise and noise with outliers.

References

1. Beck, J.V., Blackwell, B., St. Clair Jr., C.R.: Inverse Heat Conduction: Ill-Posed Problems. Wiley, New York (1985)
2. Ozisik, M.N., Orlande, H.R.B.: Inverse Heat Transfer: Fundamentals and Applications. Tailor and Francis, New York (2000)
3. Alifanov, O.M.: Inverse Heat Transfer Problems. Springer Verlag, Berlin (1994)
4. Murio, D.A.: The Mollification Method and the Numerical Solution of Ill-Posed Problems. Wiley, New York (1993)
5. Woodbury, K.A., Beck, J.V.: Estimation metrics and optimal regularization in a Tikhonov digital filter for the inverse heat conduction problem. Int. J. Heat Mass Transf. **62**, 31–39 (2013)

6. Jarny, Y., Orlande, H.R.B.: Adjoint methods. In: Orlande, H.R.B., Fudym, O., Maillet, D., Cotta, R.M. (eds.) Thermal Measurements and Inverse Techniques. CRC Press, Boca Raton (2011)
7. Borukhov, V.T.: Reconstruction of heat fluxes through differential temperature measurement by the method of inverse dynamic systems. J. Eng. Phys. Thermophys. **47**(3), 1098–1102 (1984)
8. Lee, H.-L., Chang, W.-J., Wu, S.-C., Yang, Y.-C.: An inverse problem in estimatingthe base heat flux of an annular fin based on the hyperbolic model of heat conduction. Int. Commun. Heat Mass Transf. **44**, 31–37 (2013)
9. Christov, C.I., Jordan, P.M.: Heat conduction paradox involving second sound propagation in moving media. Phys. Rev. Lett. **94**(15), 154301 (2005)
10. Papanicolaou, N.C., Christov, C.I., Jordan, P.M.: The influence of thermal relaxation on the oscillatory properties of two-gradient convection in a vertical slot. Eur. J. Mech. B. Fluids **30**, 68–75 (2011)
11. Vernotte, P.: Paradoxes in the continuous theory of the heat equation. Compt. Rend. Acad. Sci. (Paris) **246**, 3154–3155 (1958)
12. Cattaneo, C.: A form of heat conduction which eliminates the paradox of instantaneous propagation. Compt. Rend. Acad. Sci. (Paris) **247**, 431–433 (1958)
13. Luikov, A.V., Bubnov, V.A., Soloview, I.A.: On the wave solutions of heat conduction equation. Int. J. Heat Mass Transf. **19**, 245–248 (1976)
14. Chandrasekharaiah, D.S.: Hyperbolic thermoelasticity: a review of recent literature. Appl. Mech. Rev. **51**, 705–729 (1998)
15. Ván, P., Czél, B., Fülöp, T., Gyenis, G., Gróf, Á., Verha, J.: Experimenal aspects of heat conduction beyond Fourier. Thesis, 12th Joint European Thermodynamics Conference, Brescia (2013)
16. Zubair, S.M., Chaudhry, M.A.: Int. J. Heat Mass Transf. **39**(14), 3067–3074 (1996)
17. Borukhov, V.T., Kostyukova, O.I., Kurdina, M.A.: Tracking of the preset program of weighted temperatures and reconstruction of heat transfer coefficients. J. Eng. Phys. Thermophys. **83**(3), 622–631 (2010)
18. Borukhov, V.T., Kostyukova, O.I.: Identification of timedependent coefficients of heat transfer by the method of suboptimal stagebystage optimization. Int. J. Heat Mass Transf. **59**, 286–294 (2013)
19. Borukhov, V.T., Kostyukova, O.I.: Reconstruction of Heat Transfer Coefficients Using the Approach of Stage by Stage Suboptimal Optimization and Huber. Tikhonov Filtering of Input Data. Autom. Control Comput. Sci. **47**(6), 289–299 (2013)
20. Morozov, V.A.: Methods for Solving Incorrectly Posed Problems. Springer-Verlag, Heidelberg (1984)
21. Tikhonov, A.N., Arsenin, V.Y.: Solutions of Ill-Posed Problems. V.H. Winston & Sons, Washington, D.C. (1977)
22. Huber, P.J.: Robust estimation of a location parameter. Ann. Math. Stat. **35**, 73–101 (1964)
23. Huber, P.J., Ronchetti, E.M.: Robust Statistics, 2nd edn. Wiley, New York (2009)
24. Petrus, P.: Robust Huber adaptive filter. IEEE Trans. Signal Process. **47**(4), 1129–1133 (1999)
25. Binder, T., Kostina, E.: GaussNewton methods for robust parameter estimation. In: Bock, H.G., Carraro, T., Jäger, W., Körkel, S., Rannacher, R., Schlöder, J.P. (eds.) Model Based Parameter Estimation, vol. 4, pp. 55–87. Springer, Heidelberg (2013)

26. Alifanov, O.M.: Identifikatsiya protsessov teploobmena letatelnykh apparatov (vve-
denie v teoriyu obratnykh zadach) (Identification of Heat Transfer Processes of Air-
crafts (Introduction to the Theory of Inverse Problems)). Mashinostroenie, Moscow
(1979)
27. Malanowski, K., Maurer, H.: Sensitivity analysis of optimal control problems subject to higher order state constraints. Ann. Oper. Res. **101**, 43–73 (2001). (Optimization with Data Perturbations II)

Multi-Objective and Financial Portfolio Optimization of p-Persistent Carrier Sense Multiple Access Protocols with Multi-Packet Reception

Ramiro Sámano-Robles[1]([⊠]) and Atílio Gameiro[2]

[1] CISTER/INESC-TEC, ISEP, Polytechnic Institute of Porto,
Porto, Portugal
rasro@isep.ipp.pt
[2] Instituto de Telecomunicações, Campus Universitário,
3810-193 Aveiro, Portugal
amg@ua.pt

Abstract. This paper revisits the study of wireless carrier-sense multiple access (CSMA) protocols enabled with multi-packet reception (MPR) capabilities. This study employs a new paradigm in the literature of random access based on multi-objective and financial portfolio optimization tools. Under this new optimization framework, each packet transmission is regarded not only as a network resource, but also as a financial asset with different values of return and risk (or variance of the return). The objective of this network-financial optimization is to find the transmission policy that simultaneously optimizes network metrics (such as throughput and efficient power consumption), as well as economic metrics (such as fairness, return and risk). Two transmission models are considered for performance evaluation: a Bernoulli transmission model that facilitates analytic derivations, and a Markov model that considers the backlog states of the network and that facilitates dynamic stability analysis. This work is focused on the characterization of the boundary (envelope) or the Pareto optimal frontier of different types of trade-off performance region. These regions include the conventional throughput and stability regions, as well as new trade-off regions such as sum-throughput vs. fairness, sum-throughput vs. power consumption, and return vs. risk. Fairness is evaluated by means of the Gini-index, which is used in the field of economics to measure population income inequality. Transmit power is directly linked to the global transmission attempt rate. In scenarios with weak MPR capabilities, the system has problems in achieving simultaneously good values of fairness and high values of sum-throughput. This is because of an underlying non-convex throughput region which is typical of protocols dominated by unresolvable collisions. On the contrary, in scenarios with strong MPR capabilities, good fairness, higher energy consumption efficiency, and high sum-throughput performances can be simultaneously achieved. Carrier-sensing is shown to improve the convexity of the throughput region in scenarios with weak MPR, thereby achieving a better trade-off between metrics, including

© Springer International Publishing Switzerland 2015
A. Plakhov et al. (Eds.): EmC-ONS 2014, CCIS 499, pp. 68–94, 2015.
DOI: 10.1007/978-3-319-20352-2_5

return and risk. However, the effects of carrier-sensing are shown to disappear in scenarios with strong MPR capabilities or with underlying convex throughput regions. The combination of MPR with carrier-sensing tools helps in reducing risk in the network and to fight issues of wireless random access such as the hidden/exposed terminal problems.

Keywords: S-ALOHA · Random access · Multi-objective portfolio optimization · Pareto optimal trade-off curve

1   Introduction

1.1   Background, Motivations and Open Issues

The demand for wireless connectivity is rapidly increasing, particularly with the advent of the Internet-of-everything (IoE) and 5G networks. However, the scarcity of spectrum resources impedes the allocation of a dedicated channel to each device connecting to the network. New access technologies are necessary to solve this resource scarcity problem. Over the last two decades, it has been observed that large portions of licensed spectrum allocated to legacy applications remain underutilized for considerably long periods of time. This means that the dedicated spectrum assignment paradigm is obsolete. Cognitive radio (e.g., [1]), self organized networks (e.g., [2]), and software defined radio solutions attempt to provide opportunistic sharing, dynamic organization and efficient access to both licensed and unlicensed portions of spectrum. This means that future access will combine aspects and benefits of decentralized (random) and centralized (dedicated) resource allocation. This new resource sharing paradigm is expected to alleviate the issue of high spectrum demand for future applications.

Another example of the convergence of centralized and decentralized allocation can be found in the area of wireless local and personal area networks (WLANs and WPANs, respectively). The number of WLAN and WPAN hotspots has considerably increased over the last few years. This fact has raised the issue of severe interference inside buildings and in high dense urban scenarios. WLAN and WPAN decentralized technology needs to incorporate more centralized coordination algorithms due to the increasing traffic demands and higher levels of interference. It is thus clear that future wireless access networks must combine aspects of decentralized with centralized allocation to manage more efficiently network resources. This leads to the convergence of technologies such as WiFi and LTE (long term evolution) in one single solution or standard with cognitive radio and self organized features. This framework highlights the importance of the study of random access protocols using modern optimization and signal processing tools compatible with the literature of centralized networks. This is in view of future synergies and the potential convergence of both domains.

Random access protocols represent one of the cornerstones of any wireless multi-user communication system. In centralized networks such as UMTS (universal mobile telecommunication systems) and LTE, random access protocols are used whenever terminals request initial access to network resources. In wireless local and personal area networks, random access protocols are the core of the dominant technology standards (e.g., IEEE 802.1 and IEEE 802.15.4). The ALOHA protocol is the text book example of theory of random access. Since its proposal in the seminal work of Abramson in [3], ALOHA has been target of multiple reinterpretations and improvements. Recent approaches have revisited the analysis of ALOHA with advanced schemes such as multi-packet reception (MPR) [4], cooperative diversity [5], and multi-hop ad-hoc features [6]. Perhaps the most significant, effective and widely implemented variation of ALOHA is the carrier-sense multiple access (CSMA) protocol. In CSMA, terminals with a packet ready to be transmitted sense the channel before deciding to engage in transmission [7]. The gain provided by CSMA depends on how often the sensing operation is performed along the duration of a packet transmission. This carrier-sensing scheme is efficient in wire-line solutions, evolving to the current Ethernet IEEE 802.3 standards, which also employ collision detection mechanisms to further improve channel utilization. However, in wireless settings, CSMA is affected by the hidden and exposed terminal problems. Practical solutions implemented by current WLAN and WPAN standards to counteract this impairment include collision avoidance and resource reservation schemes. Another potential solution to the hidden/exposed terminal problems is the use of multi-packet reception capabilities[1] via multiple reception antennas (e.g. [8]) and retransmission diversity (e.g. [9]). Even when terminals incur in errors of carrier-sensing due to hidden or exposed settings, collisions can still be resolved by means of a strong physical (PHY) layer.

ALOHA and its carrier-sense version have been mainly subject to conventional single objective optimization approaches (e.g., [7,10]). To the best of our knowledge there are no previous works that address the *multi-objective optimization* of these protocols and in general in the field of random access. In addition to this, the use of multiple-input multiple-output (MIMO) tools, which is widespread in the literature of centralized networks, is not as rich and diverse in the field of random or decentralized resource allocation. This gap needs to be filled in view of a future convergence of the fields of random and dedicated resource allocation using multiple antennas or multi-packet reception. This paper attempts to partially address these issues as explained in more detail in the following sections.

1.2   Paper Objectives and Contributions

This paper addresses the multi-objective and financial portfolio optimization of a p-persistent CSMA protocol with MPR capabilities. To achieve this goal we use the conditional probabilistic reception model proposed in [4] and the p-persistent transmission model presented in [11]. The analysis assumes fixed-length packets.

---

[1] Multi-packet reception is the ability of the PHY-layer to correctly decode concurrent or contending transmissions, mainly by using signal processing tools for multiple-input multiple-output (MIMO) systems.

The derivation of the boundary (envelope) of the throughput region is reformulated here as the simultaneous optimization of the individual terminal throughput functions. The Pareto optimal curve (surface) is identified as the envelope of the throughput region. The remaining trade-off regions analysed are the following: stability region, sum-throughput vs. fairness, sum-throughput vs. power, and return vs. risk regions. Power consumption is measured as the total transmission attempt rate, which is an assumption commonly used in the study of energy consumption of random access protocols (e.g. [11]). Fairness is evaluated in this paper by means of the Gini-index, which is used in economics to measure income inequality [12,13]. Finally, the characterization of the return vs. risk trade-off region employs concepts borrowed from the theory of financial portfolio optimization (see [14]). Each network transmission will be also considered as a financial asset, whose allocation will attempt to simultaneously maximize return and minimize risk (variance of the return).

Economic optimization tools have been widely used in wireless networks, using, for example, financial stock market tools and game theory. However, these works have been mainly used in cellular resource allocation, cognitive radio networks, and for operator price estimation scenarios. This work attempts to pioneer the use of financial portfolio optimization in the particular case of random access. Summarizing, the contributions of this paper are as follows:

1. Multi-objective optimization of a p-persistent CSMA-MPR protocol, which allows for a trade-off analysis of different objective functions.
2. Derivation of Pareto optimal front curves for different performance trade-off regions: the conventional throughput and stability regions, sum-throughput vs. power, sum-throughput vs. fairness, and return vs. risk regions.
3. Geometric interpretation of some Pareto optimal trade-off curves, and
4. Innovative use of financial terms (return and risk) in the context of random access networks.

1.3   Related Works

The ALOHA protocol was originally proposed in the seminal work of Abramson in [3]. With a relatively poor performance in terms of channel efficiency (18 %), improvements were soon proposed using slotted transmission [15] and carrier-sensing (CSMA) [10]. The protocol was also found to be inherently unstable (or bistable) [16]. Since then, improved dynamic stability analysis and stabilization schemes have been proposed (e.g., [7]). The power capture effect has been proved to be helpful in stabilization of ALOHA [17,18]. The first cross-layer optimization approach for the ALOHA protocol with MPR was provided in [19]. A stochastic MPR matrix was used in the sum-throughput optimization of a symmetrical system using an infinite population model. Dynamic allocation schemes for this MPR model were proposed in [20–22]. An extension to the asymmetrical case was provided in [4] using a conditional probabilistic reception model and a finite buffered population.

Optimization of other types of symmetrical S-ALOHA systems can be found in [5] for systems with cooperative diversity, in [7,10,11] for CSMA protocols, and in [8,19] for systems with MPR. More recently, the works in [23,24] have addressed the optimization with decentralized channel state information (CSI). Optimization of S-ALOHA with MPR using game theory was presented in [25]. The present work is complementary to these recent approaches. The aim of this paper is to explore all the range of Pareto optimal solutions (i.e., Pareto frontier) and the different trade-offs between Pareto solutions. This full trade-off analysis provides a useful engineering design perspective of wireless networks and leads to a better understanding of complex systems and the different variables involved in their design. By contrast, game theory only searches for solutions that comply with the Nash equilibrium condition, thus representing a more limited approach.

In terms of asymmetrical settings[2], the literature of random access is relatively scarce. It is known that closed-form expressions for the optimization of certain random access protocols and networks in asymmetrical settings only exist for limited number of terminals, in general only for $J = 2$ and a few cases with $J = 3$ terminals. The stability region of ALOHA with MPR has been proved to be identical to the throughput region for the case of $J = 2$ in [26]. The exact non-parametric closed-form expression of the stability region for a two-user ALOHA-MPR system was derived in [4][3]. Approximate stability conditions for higher numbers of users ($J > 2$) are given in [4] for the MPR channel and in [27] for the conventional ALOHA using a queuing rank analysis. The optimum transmission policy for conventional ALOHA that characterizes the throughput region has been derived in [16], and in [5] for CSMA protocols assisted by cooperative diversity. The throughput region of a p-persistent CSMA protocol with only two terminals has been derived analytically in [28]. However, closed-form expressions for other systems remain elusive in the literature, particularly when considering an arbitrary numbers of terminals.

The optimization approach used in this work addresses fully asymmetrical settings focusing on the derivation of complex Pareto front curves for arbitrary numbers of terminals. However, for convenience, this paper addresses in more detail the particular case of two terminals, for which closed-form expressions and exact sketches of the different regions are relatively easier to obtain and which provide an idea of the main relationships between metrics. Based on the results for two-user systems, the case with more than two terminals is addressed by approximate analytic expressions that can be solved numerically or which can be explained using simplified sketches of the different trade-off regions under investigation.

Regarding techno-economic analysis of wireless networks, several works exist in the literature. The conventional approach is the use of a techno-economic performance model to evaluate the revenue of an operator under a given set of resource allocation assumptions. The main objective is to find the optimum

---

[2] Asymmetrical settings is used to refer to network models where users are explicitly modelled with different channel and queuing statistics.

[3] Stability region is loosely defined here as the set of arrival rates for which the queues of all users remain bounded or empty within an finite period of time. Throughput region can also be loosely defined as the set of achievable throughput terminal values.

resource allocation that provides the highest revenue and that satisfies the users of the network [29]. In the context of cognitive radio, research efforts have been intensive over the last ten years due to the relevance of the understanding of the potential gains of opportunistic spectrum access. A review of different approaches for the use of economic optimization tools in cognitive radio can be found in [30], where the authors have proposed a market equilibrium approach where primary and secondary users implement a learning algorithm so that they can adapt the amount of spectrum used, their pricing and the optimum demand that achieve equilibrium. Most of the existing works are based on game theoretic concepts (see [31–35]). The work in [34] has used an atomic congestion game theoretic approach in a wireless network with spatial reuse and inter-user interference. The work in [35] addresses the problem of calculating the optimum spectrum pricing in a dynamic spectrum market. Another related approach for the use of economics in cognitive radio can be found in works such as [36,37] and references therein, which are based on the concepts of auction theory.

This paper uses multi-objective portfolio optimization under the assumption that each packet transmission is also a financial asset. Our work explicitly introduces the concept of risk in the resource allocation problem of random access and derives relevant expressions that allow for an interpretation of the resource allocation problem as a financial stock market. The work in [38] has used the concept of return and variance of the return in the context of spectrum pricing and copyright. Our approach is different from these previous works regarding the explicit use of multi-objective optimization and the exploration of the boundaries of different Pareto optimal trade-off curves. This allows us to visualize geometrical attributes and the potential trade-offs between network and economic performance metrics. In other words, instead of deriving a resource allocation policy that achieves a Nash or market equilibrium as in previous works, here we explicitly explore the boundaries of different trade-off performance regions or the Pareto frontier curve (surface). In this sense, our approach complements previous works in the literature by providing a framework for trade-off analysis and explicit interpretation of financial market stock tools in wireless networks.

Strictly speaking the framework presented in this paper can be used for higher dimensionality Pareto frontier analysis. However, the trade-off regions analysed are two-dimensional, as this facilitates analysis via sketches that provide more useful information than a higher dimensional representation. All the results provide also the projections of the different trade-off regions, which provides all the necessary information of the remaining dimensions of the Pareto frontier.

1.4 Paper Organization

The structure of this paper is as follows. Section 2 describes the system model. Section 3 defines the trade-off performance regions to be investigated. Section 4 addresses the multi-objective optimization. Section 6 presents sketches of the different trade-off Pareto optimal front curves, and finally Sect. 7 presents the conclusions.

2   System Model

2.1   Scenario Description

Consider the slotted random-access network depicted in Fig. 1 with one base station (BS) and J user terminals. The BS is provided with multiple antennas which enable MPR, defined as the ability of the physical layer to correctly receive concurrent transmissions. Users have only one antenna and are assumed to have their own buffer with incoming packets always available to be transmitted (dominant system or full-queue assumption). At the beginning of every time-slot, each user j will sense the channel. If the result of this sensing operation is idle, then the terminal will start a random packet transmission process (see next subsection for the description of this randomized transmission process). Multi-packet reception is evaluated using the conditional probabilistic reception model proposed in [4]. The authors have defined the marginal probabilities of reception for user j, conditional on the transmission of a set of active users $(\mathcal{T})$ as follows:

$$q_{j|\mathcal{T}} = \sum_{j \in \mathcal{R}} q_{\mathcal{R};\mathcal{T}}, \qquad \mathcal{R} \subset \mathcal{T}, \qquad j \in \mathcal{T}, \tag{1}$$

where $q_{\mathcal{R};\mathcal{T}}$ is the probability of decoding packets *only* from the set of users $\mathcal{R}$ conditional on the set of transmitting users $\mathcal{T}$. This marginal conditional reception model describes several statistical features found in wireless networks, particularly their *asymmetric* nature. Now suppose that $\mathcal{U}$, $\mathcal{S}$, and $\hat{\mathcal{S}}$ are three groups of terminals. The MPR channel is standard when the following inequality holds for all $\mathcal{U} \subset \mathcal{S} \subset \hat{\mathcal{S}}$ [26]:

$$\sum_{\mathcal{R};\mathcal{U} \subset \mathcal{R} \subset \hat{\mathcal{S}}} q_{\mathcal{R};\mathcal{S}} \geq \sum_{\mathcal{R};\mathcal{U} \subset \mathcal{R} \subset \hat{\mathcal{S}}} q_{\mathcal{R};\hat{\mathcal{S}}}.$$

This condition states that collisions of higher numbers of users will be always more destructive and thus less likely to be resolved than those collisions with less users. A packet is assumed to have a fixed length (in time-slots or packet units) denoted by L, which also denotes the number of times the sensing operation is performed along the duration of a packet. For convenience in the analysis, the random variable l will denote the length of a transmission or renewal interval [11]. Two transmission models will be used for the study of the protocol that will help in revealing different aspects of the system, and which are described in the following subsections.

2.2   Bernoulli Transmission Model

In the Bernoulli transmission model, all traffic streams either backlogged[4] or new incoming are not differentiated. Therefore, at the beginning of every time-slot and provided the channel was sensed as idle, each user j will be assumed

---

[4] A user is said to be in the backlog state when having previously transmitted a packet, the transmission was lost in a collision and the packet needs to be re-transmitted in subsequent time-slots.

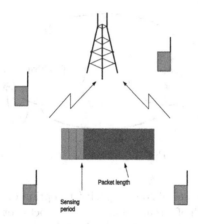

Fig. 1. Random access network with MPR capabilities and carrier-sensing.

to attempt a packet transmission controlled by a Bernoulli random experiment with parameter p_j, which is also the transmission probability. The advantage of the Bernoulli transmission model is that it facilitates analytic derivations. However, backlog and incoming traffic streams are not differentiated, and thus it is not possible to evaluate in detail the dynamics and stability properties of the protocol.

2.3 Dynamic State Model

To overcome the limitations of the Bernoulli transmission model, the operation of the protocol can be reformulated to consider incoming and backlog traffic being scheduled in different manner. Each terminal is assumed to be in two possible states (see Fig. 2): *idle* (with probability $p_{i,j}$), or *backlog* (with probability $p_{b,j}$). In the idle state, a terminal attempts the transmission of a new incoming packet with probability $p_{a,j}$. In the case of collision and upon the reception of the feedback signal from the BS confirming the collision event, each user is driven into the backlog state. In the backlog state, each user will attempt the re-transmission of the packet previously lost with probability $p_{r,j}$. Not new incoming traffic is allowed to be transmitted by any user while being in the backlog state.

3   Trade-off Performance Regions

3.1   Throughput Region

Under CSMA operation, the throughput of each terminal is given by the ratio of the average number of correctly received packet units to the average length of a renewal interval [11]. In our setting, this can be expressed as follows:

$$T_j = \frac{L p_{s,j}}{E[l]}, \tag{2}$$

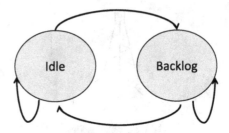

Fig. 2. Terminal states.

where $p_{s,j}$ is the probability of correct packet reception of a packet of user j and $E[l]$ is the average length of a renewal interval or epoch-slot ($E[\cdot]$ is the statistical average operator). Using the reception model defined in (1), the term $p_{s,j}$ in the numerator of (2) can be calculated as the expectation over the probability space of all possible realizations of the set of contending users $\mathcal{T}$ that include user j ($j \in \mathcal{T}$):

$$p_{s,j} = \sum_{\mathcal{T}:j\in\mathcal{T}} \Pr\{\mathcal{T}\}q_{j|\mathcal{T}}, \qquad j \in \mathcal{T}, \tag{3}$$

where $\Pr\{\mathcal{T}\}$ indicates the probability of occurrence of a particular set of transmitting users $\mathcal{T}$, and which can be written, under the assumption of independent queues[5], as follows

$$\Pr\{\mathcal{T}\} = \prod_{i\in\mathcal{T}} p_i \prod_{j\notin\mathcal{T}} \bar{p}_j,$$

where $\bar{a} = 1 - a$, for any a. We can also obtain an expression for the average length of an epoch $E[l]$ by using the formula for a finite-user CSMA protocol with constant packet length [11]:

$$E[l] = L(1 - \Pr\{\mathcal{T} = \emptyset\}) + \Pr\{\mathcal{T} = \emptyset\},$$

where $\Pr\{\mathcal{T} = \emptyset\}$ is the probability that no user transmits, i.e. the probability that the set of of transmitting users is empty. This expression simply denotes that the length of the epoch is L whenever there is a packet transmission with probability $1 - \Pr\{\mathcal{T} = \emptyset\}$ plus the contribution of only one time-slot whenever there is no user transmitting with probability $\Pr\{\mathcal{T} = \emptyset\}$. Now, let us rewrite the above expression for $E[l]$ as follows:

$$E[l] = L + (1 - L)\Pr\{\mathcal{T} = \emptyset\},$$

and, assuming again independence of queues, it finally reduces to:

$$E[l] = L + \bar{L}\prod_{j=1}^{J} \bar{p}_j \tag{4}$$

---

[5] Queues in a random access in general are not statistically independent, particularly in the presence of collisions. However, at low and medium traffic loads it is a good approximation commonly used in the literature [9,16].

Having defined the throughput expressions per terminal, let us now turn our attention to the throughput region of the protocol. Consider the vector $\mathbf{T} = [T_1, T_2, \quad \dots \quad T_J]^T$ of stacked throughput values, and the vector $\mathbf{p} = [p_1, p_2, \quad \dots p_J]^T$ of stacked transmission probabilities. The *throughput region*, can be defined as the union of all achievable values $[T_1, T_2, \quad \dots T_J]$ for all potential realizations of transmission policies $(0 < p_j < 1)$ [26]:

$$\mathcal{C}_T = \{\tilde{\mathbf{T}} | \tilde{T}_j = T_j(\mathbf{p}), 0 \le p_j \le 1\}. \tag{5}$$

The throughput region is one of the main metrics in the study of random access in asymmetrical settings [26].

3.2   Sum-throughput vs. Fairness Region

Considering the individual throughput expressions in (2), sum-throughput can be simply defined as follows:

$$T = \sum_{j=1}^{J} T_j. \tag{6}$$

Fairness will be evaluated in this paper by means of the Gini-index, which is commonly used in the area of economics to measure income inequality [13]. The Gini-index can be defined mathematically as follows [13]:

$$F_G = \frac{\sum_{j=1}^{J} \sum_{k=1}^{J} |T_j - T_k|}{2J^2\mu} = \frac{\sum_{j=1}^{J} \sum_{k=1}^{J} a_{j,k}(T_j - T_k)}{2JT}, \tag{7}$$

where $\mu = \sum_{j=1}^{J} T_j / J$ is the mean, and $a_{j,k}$ is defined as $a_{j,k} = \begin{cases} 1, & T_j \ge T_k \\ -1, & T_j < T_k \end{cases}$.
A value of Gini-index of zero $(F_G = 0)$ is equivalent to the maximum fairness case where all users are statistically identical. On the contrary, a value of one $(F_G = 1)$ indicates the worst fairness scenario with one user overtaking all the resources of the system. Consider the vector $\mathbf{F} = [T \quad F_G]^T$ of stacked values of sum-throughput and fairness. The sum-throughput vs. fairness trade-off region can be defined as the union of all achievable values $[T \quad F_G]^T$ for all potential realizations of transmission policies $(0 < p_j < 1)$:

$$\mathcal{C}_F = \{\tilde{\mathbf{F}} | \tilde{T} = T(\mathbf{p}), \tilde{F}_G = F_G(\mathbf{p}), 0 \le p_j \le 1\}. \tag{8}$$

3.3   Sum-throughput vs. Transmit Power Region

In this paper, average power consumption will be considered as proportional to the transmission attempt rate of the system, which is a common assumption used in the literature of random access (i.e. [11]). Therefore, we can define the average consumed power as follows:

$$P = \alpha \sum_{j=1}^{J} p_j, \tag{9}$$

where α is a proportionality constant that relates each transmission with a specific energy consumption. Consider the vector $\mathbf{P} = [T \quad P]^T$ of stacked values of sum-throughput and power. The sum-throughput vs. power trade-off region can be defined as the union of all achievable values $[T \quad P]^T$ for all potential realizations of transmission policies $(0 < p_j < 1)$:

$$\mathcal{C}_P = \{\tilde{\mathbf{P}} | \tilde{T} = T(\mathbf{p}), \tilde{P} = P(\mathbf{p}), 0 \leq p_j \leq 1\}. \tag{10}$$

3.4   Return vs. Risk Trade-off Region

Let us define the instantaneous return per correctly transmitted packet of user j as r_j, and the average return as $E[r_j] = \hat{r}_j$. The average return of the network can be written as:

$$R = \sum_{j=1}^{J} E[r_j t_j] = \sum_{j=1}^{J} \hat{r}_j T_j, \tag{11}$$

where t_j is a binary random variable that indicates whether the packet of user j was correctly received $(t_j = 1)$ or not $(t_j = 0)$. Note that because t_j is a binary random variable, $E[t_j] = E[t_j^2] = T_j$. In this paper we consider that the return of different users is statistically independent. The risk is defined the variance of the instantaneous return:

$$S = E[(\sum_{j=1}^{J} r_j t_j)^2] - R^2 = \sum_{j=1}^{J} E[r_j^2] T_j - R^2. \tag{12}$$

Consider the vector $\mathbf{R} = [R \quad S]^T$ of stacked values of return and risk. The *return vs. risk trade-off region* can be defined as the union of all achievable values $[R \quad S]^T$ for all potential realizations of transmission policies $(0 < p_j < 1)$:

$$\mathcal{C}_R = \{\tilde{\mathbf{R}} | \tilde{R} = R(\mathbf{p}), \tilde{S} = S(\mathbf{p}), 0 \leq p_j \leq 1\}. \tag{13}$$

4   Optimization

4.1   Multi-objective Optimization

To derive the envelope of the different trade-off regions, a multi-objective optimization scheme is here proposed, where M objective functions F_m ($m = 1, \ldots, M$) can be simultaneously optimized:

$$\mathbf{P}_{opt} = \arg\max_{\mathbf{P}} \ [F_1, F_2 \quad \cdots \quad F_M], \qquad 0 < p_j < 1. \tag{14}$$

Since this vector optimization usually lacks a unique solution [39], the concept of Pareto optimal trade-off front is commonly employed. A Pareto optimal solution can be loosely defined here as the point that is at least optimum for one or more of the elements of the vector objective function $[F_1, F_2 \quad \cdots \quad F_M]$, or in other words when none of the objective functions can be improved in value without degrading some of the other objective values (see [39] for a complete definition). The multi-objective optimization problem can be transformed into a single objective optimization problem using the method of scalarization [39]:

$$\mathbf{P}_{opt} = \arg\max_{\mathbf{P}} \sum_{m=1}^{M} \mu_m F_m, \qquad 0 < p_j < 1, \tag{15}$$

where μ_m is the relative weight given to the mth objective function. Differentiating the objective function in (15) we obtain a set of equations given by $\sum_{m=1}^{M} \mu_m \frac{\partial F_m}{\partial p_k} = 0$, $k = 1.., J$. The solution of this set of linear equations independent from the values of the weighting factors μ_k can be easily proved, in our context, to be equivalent to setting the following Jacobian determinant equal to zero [9]:

$$|\mathbf{J}_a| = 0, \qquad 0 < p_j < 1, \tag{16}$$

where $J_a(m, k) = \frac{\partial F_m}{\partial p_k}$ is the (m, k) element of the Jacobian matrix $\mathbf{J}_a$.

4.2   Throughput Region

In the case of the throughput region, the M objective functions to be optimized are given by the throughput functions T_j of each user. Let us expand the term $p_{s,j}$ in (3) explicitly in terms of the transmission probabilities p_j as follows:

$$p_{s,j} = p_j(q_{j|\{j\}} + \sum_{k=1}^{J-1} \sum_{T_k, j \notin T} \prod_{i \in T} p_i Q_{j,T_k}) \tag{17}$$

where:

$$Q_{j,T_k} = q_{j|\{j\}} + \sum_{n=1}^{k}(-1)^n \sum_{\hat{T}_n \subseteq T_k} q_{j|\hat{T}_n}, \qquad j \notin T_k, \tag{18}$$

and T_n and $\hat{T}_n$ denote sets of n terminals ($|T_n| = n$, where $|\mathcal{U}|$ denotes the cardinality of $\mathcal{U}$). The optimum transmission policy can be obtained by solving the Jacobian determinant equation in (16) using the expressions of throughput and the expressions in (17) and (18). A closed-form solution of this problem is in general difficult to obtain. This paper proposes a method that provides a solution in closed-form by considering that the desired solution is a deviation from the solution of an equivalent collision model protocol. The solution for the optimization of the throughput region of random access protocols under the collision model results in Jacobian $\tilde{\mathbf{J}}_a$ matrices (following the lines of the derivation of the expression in (16)) that have the following quasi-symmetrical property:

$$\tilde{J}_a(j, k) = \begin{cases} \alpha_j, k = j \\ \beta_j, k \neq j \end{cases} \tag{19}$$

which means that all the elements of a row j are all the same except for the element of the main diagonal. Under this structure, the Jacobian determinant $|\tilde{\mathbf{J}}_a|$ has been proved in ([9]) to be equal to

$$|\tilde{\mathbf{J}}_a| = 1 - \sum_j \left\{ \frac{\beta_j}{\alpha_j - \beta_j} \right\}.$$

The structure of the Jacobian matrix for the MPR case in general does not have the same structure as in the case of the collision model protocols. However, the elements can be arranged in a way that is quasi-symmetrical or slightly approximate to a collision model matrix, and then propose complement that produces the desired quasi-symmetrical property. This can be mathematically expressed as follows:

$$\tilde{J}_a(j, k) = J_a(j, k) + \dot{J}_a(j, k),$$

where $\dot{J}_a(j, k)$ is the element of Jacobian matrix $\dot{\mathbf{J}}_a$ that complements the original Jacobian matrix $\mathbf{J}_a$ to acquire the desired symmetrical property defined in (19). The Jacobian determinant can be now calculated (using the well known co-factors formula) as the determinant of the symmetrical collision model matrix component $\tilde{\mathbf{J}}_a$ minus the deviation component that can be obtained by analysing each one of the components (co-factors) of the complement matrix $\dot{\mathbf{J}}_a$. This can be mathematically expressed as follows:

$$|\mathbf{J}_a| = |\tilde{\mathbf{J}}_a| - \sum_{j=1}^{J} (-1)^j \{ J_a(1, j)(|\tilde{\mathbf{J}}_a^{1,j}| - |\mathbf{J}_a^{1,j}|) + \dot{J}_a(1, j)|\tilde{\mathbf{J}}_a^{1,j}|\}, \qquad (20)$$

where $\mathbf{B}^{k,j}$ denotes the submatrix that is formed by removing the k-th row and the jth column of matrix $\mathbf{B}$. Details of the derivation of this formula have been omitted as they are out of the general scope of this paper. In the case of two users, this expression can be proved to reduce to [8]:

$$1 - p_1 Q_{2|\{1\}}/q_{2|\{2\}} - p_2 Q_{1|\{2\}}/q_{1|\{1\}} - \bar{L}\bar{p}_1\bar{p}_2\bar{Q} = 0, \qquad (21)$$

where $\bar{Q} = 1 - Q_{1|\{2\}}/q_{1|\{1\}} - Q_{2|\{1\}}/q_{2|\{2\}}$. Note that when using $L = 1$ in the previous expression, the expression is identical to the result for the stability region derived in [4] for the ALOHA protocol with MPR capabilities $(p_1 Q_{2|\{1\}}/q_{2|\{2\}} + p_2 Q_{1|\{2\}}/q_{1|\{1\}} = 1)$.

4.3   Sum-throughput vs. Fairness Region

Consider that the two objective functions to be optimized are $F_1 = F_G$ in (7) and $F_2 = T$ in (6). For the multi-user case, the solution boils down to a set of Jacobian determinants using the general expression in (16) for all combinations of two terminals. The dimension of each determinant matrix is 2×2. As it wil be observed later, this set of expressions describes several frontiers in the solution space. Only one of these solutions will represent the desired Pareto frontier, which is in general the part of the solution corresponding to the user with the best reception performance. This will be explained in more detail later in the section of results as a consequence of the analysis of the solution of the two-user case. In the case of two users, the Jacobian determinant equation in (16) reduces to:

$$1 - p_1 Q_{2|\{1\}}/q_{2|\{2\}} - p_2 Q_{1|\{2\}}/q_{1|\{1\}} - \bar{L}\bar{p}_1\bar{p}_2\bar{Q} = 0, \tag{22}$$

which is identical to the solution for the throughput region in (21). Therefore, the Pareto frontier in both trade-off regions is described by the same functional expression.

4.4 Sum-throughput vs. Transmit Power Region

Consider that the two objective functions to be optimized are now given by $F_1 = P$ in (9) and $F_2 = T$ in (6). Similar to the previous case of the fairness region, the solution is described by a set of Jacobian determinant equations for all combinations of two terminals. The desired frontier will be described by the part of the solution corresponding to the user with the best performance in terms of throughput, as it will be analysed in detail in subsequent sections. For the two user case, the expression in (16) becomes:

$$\frac{\partial T}{\partial p_1} - \frac{\partial T}{\partial p_2} = 0, \tag{23}$$

which after substituting the expression for T in (6) and a few algebraic operations can be rewritten as follows:

$$L[q_{1|\{1\}} + p_1(Q_{1|\{2\}} + Q_{2|\{1\}})] + \bar{L}\bar{p}_2 + L[q_{2|\{2\}} + p_2(Q_{1|\{2\}} + Q_{2|\{1\}})] + \bar{L}\bar{p}_1 = 0. \tag{24}$$

Further details are given in the section of results.

4.5 Return vs. Risk

Consider that the two objective functions to be optimized are given by $F_1 = R$ in (11) and $F_2 = S$ in (12). The analysis of the multi-user case is similar to the two previous trade-off regions. For the two user case, (16) becomes

$$\frac{\partial R}{\partial p_1}\frac{\partial S}{\partial p_2} = \frac{\partial R}{\partial p_2}\frac{\partial S}{\partial p_1}. \tag{25}$$

Simplifying and using the explicit formula for return and risk in (11) and (12), respectively, (25) can be proved to reduce to:

$$1 - p_1 Q_{2|\{1\}}/q_{2|\{2\}} - p_2 Q_{1|\{2\}}/q_{1|\{1\}} - \bar{L}\bar{p}_1\bar{p}_2\bar{Q} = 0, \tag{26}$$

which is identical to the solution for the cases of throughput and sum-throughput vs. fairness regions in (21) and (22), respectively. It is worth pointing out that despite the equivalence of the equations that describe the Pareto front curve of different trade-off regions, their actual shape and interpretation can actually differ in some aspects that will be discussed in the section of results.

5 Markov Model

For a more detailed analysis of the stability properties of the protocol, this section makes clear distinction between incoming and backlog traffic streams. Each user is assumed to be in one of two states: *idle* or *backlog*. All the J^2 possible backlog states of the network are rearranged in a linear array. The vector of backlog state probabilities in the steady state is denoted here by $\mathbf{x} = [x_1, \ldots, x_{J^2}]^T$, where x_k is the probability of the network being in the backlog state k. The transition probabilities between the different states of the network can be arranged in a $J^2 \times J^2$ matrix denoted here by $\mathbf{M}_e$. Therefore, the element k, l of matrix $\mathbf{M}_e$, denoted by $M_e(k, l)$, indicates the transition probability between the lth and the kth state. The transition probabilities can be obtained by simply adding the probabilities of possible conflict or no conflict between users. This can be written as the transition probability between the backlog state of the network at time t and the backlog state at time $t + 1$. The backlog state of the network at time t is defined as the set of users in backlog state at time t, which is denoted by $\mathcal{U}_t$. The transition probability between the two consecutive states is defined as the probability of occurrence of the set of backlogged users $\mathcal{U}_{t+1}$ at time-slot $t + 1$ conditional on the occurrence of the set of backlogged users $\mathcal{U}_t$ in the previous time-slot t. This can be written as:

$$\Pr\{\mathcal{U}_{t+1}|\mathcal{U}_t\} = \sum_{\mathcal{T}_a, \mathcal{T}_b} \Pr\{\mathcal{T}_a; \mathcal{T}_b\} q_{\mathcal{R};\mathcal{T}}, \quad \mathcal{T}_b \subset \mathcal{U}_t, \quad \mathcal{T}_a \subset \overline{\mathcal{U}_t}, \tag{27}$$

where $\mathcal{T}_a$ and $\mathcal{T}_b$ are the subsets of users in idle and backlog states, respectively, that engage in a packet transmission, $\mathcal{R} = \overline{\mathcal{U}_{t+1}} \cap \mathcal{T} \cap \mathcal{T}$ is the subset of active users correctly decoded, $\overline{\mathcal{S}}$ denotes the complement of set $\mathcal{S}$, and $\mathcal{T} = \mathcal{T}_a \cup \mathcal{T}_b$ is the total set of users engaged in transmission. The term $\Pr\{\mathcal{T}_a; \mathcal{T}_b\}$ denotes the probability of occurrence of the set of active users $\mathcal{T}_a$ and backlogged users $\mathcal{T}_b$ engaged in transmission. The steady state probabilities can be obtained by solving the characteristic equation of the Markov model:

$$\mathbf{M}_e \mathbf{x} = \mathbf{x}, \tag{28}$$

which can be identified as an eigenvalue problem with a particular eigenvalue equal to one ($\nu = 1$). Throughput functions can be reformulated by averaging the contributions over the calculated backlog probability space as follows:

$$T_j = \sum_{\mathcal{U}} \Pr\{\mathcal{U}\} \sum_{\mathcal{T}_a; \mathcal{T}_b} \Pr\{\mathcal{T}_a; \mathcal{T}_b\} q_{j|\mathcal{T}}, \quad \mathcal{T} = \mathcal{T}_a \cup \mathcal{T}_b, \quad j \in \mathcal{T},$$

where the summation is over all possible sets of users in backlog state and over all possible combinations of users engaged in transmission either in the idle ($\mathcal{T}_a$) or in the backlog states ($\mathcal{T}_b$). Similarly, the average power consumption is given by:

$$P = \sum_{\mathcal{U}} \Pr\{\mathcal{U}\} \sum_{\mathcal{T}_a; \mathcal{T}_b} \Pr\{\mathcal{T}_a; \mathcal{T}_b\} \sum_{j \in \mathcal{T}_a, k \in \mathcal{T}_b} (p_{a,j} + p_{b,k})$$

The probabilities of each user being in the idle state can be also obtained by averaging over the calculated probability space:

$$p_{b,j} = \sum_{\mathcal{U},j\in\mathcal{U}} \Pr\{\mathcal{U}\} \tag{29}$$

while the offered load of each user can be written as the average accepted traffic:

$$\lambda_j = \sum_{\mathcal{U};j\notin\mathcal{U}} \Pr\{\mathcal{U}\}p_{a,j}E[l] \tag{30}$$

The expressions for the sum-throughput, fairness indicator, return and risk remain as in the previous section, except that they must be calculated as the average over the probability space obtained using the Markov model tools.

5.1   Stability Region

Let us now define stability region. For the sake of subsequent calculations, let $\overrightarrow{\lambda} = [\lambda_1, \lambda_2 \ldots, \lambda_J]^T$ be the vector of stacked arrival rate values of all terminals. The stability region C_λ is the union over all possible realizations of arrival rates for which the output traffic is larger than the backlog and incoming traffic streams for all users:

$$C_\lambda = \{\overrightarrow{\lambda}\,|\,\tilde{\lambda}_j, p_{a,j}p_{i,j} + p_{r,j}p_{b,j} \leq p_{s,j}\}. \tag{31}$$

The boundaries of the stability region can be obtained by means of a multi-objective optimization in terms of the probabilities of transmission in idle and backlog states. The objective is to find the operational points where the balance between incoming, backlog and outgoing traffic is achieved. This defines the Pareto frontier or the boundary of the stability region.

6   Results and Discussion

This section presents sketches of the different trade-off performance regions. Each figure is dedicated to one of the trade-off regions. For purposes of comparison, each figure also includes the projections of the other types of trade-off region being analysed. Two cases for the selection of the reception parameters are here presented with weak and strong MPR capabilities. The reception parameters for the weak MPR case are: $q_1 = 0.9$, $q_2 = 0.7$, $q_{1|\{1,2\}} = 0.1$, and $q_{2|\{1,2\}} = 0.1$. In the case of strong MPR the following parameters have been used: $q_1 = 0.9$, $q_2 = 0.7$, $q_{1|\{1,2\}} = 0.6$, and $q_{2|\{1,2\}} = 0.4$.

6.1   Throughput Region

Figures 3, 4 and 5 present the sketches of the throughput region. Figures 3 and 4 show the results for weak MPR using, respectively, $L = 1$ (which is equivalent to an ALOHA system without carrier-sensing), and $L = 8$ which is the CSMA system. Figure 5 shows the results with strong MPR. The envelope of these regions is given by boundary conditions and by the expression for the optimum transmission probabilities in (21). It can be observed that in the case of weak MPR (Figs. 3 and 4) the throughput region has a non-convex shape. Note that by using carrier sensing the throughput region reduces its non-convexity, while the use of strong MPR produces convex region due to improved reception conditions. In the strong MPR case (Fig. 5), the results are identical for different values of packet length L. This means that effects of carrier-sensing disappear. The top envelope of the throughput region in the weak MPR case[6] has a three sections given by boundary conditions and the curve tagged as $\frac{p_1 Q_2}{q_2} + \frac{p_2 Q_1}{q_1} = 1 - \bar{p}_1 \bar{p}_2 \bar{Q}$. The strong MPR case has only two sections given only by boundary conditions ($p_1 = 1$ or $p_2 = 1$). The equal throughput line at 45 degrees is tagged as $T_1 = T_2$.

6.2   Sum-throughput vs. Fairness Region

Figures 6, 7 and 8 present the sketches of the $\mathcal{C}_F$ region in (8). Figures 6 and 7 show the results for weak MPR using, respectively, $L = 1$ and $L = 8$. Figure 8 shows the results with strong MPR. The envelope is given by boundary conditions and by the expression for the optimum transmission probabilities in (22). The $\mathcal{C}_F$ regions are shown to be upper bounded by one portion of the curve that defines the upper bound of the throughput region, which corresponds to the user with better reception capabilities. This property extends to the multi-user case too. The point with maximum sum-throughput with the worst Gini-index at the

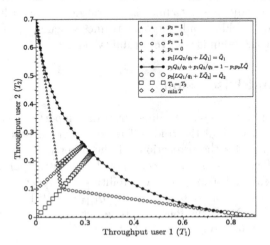

Fig. 3. Throughput region (weak MPR) with $L = 1$ (ALOHA).

---

[6] Strong MPR is defined as the scenario where the throughput region becomes convex. The exact definition can be found in [19].

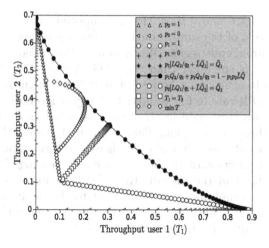

Fig. 4. Throughput region (weak MPR) with $L = 8$ (CSMA).

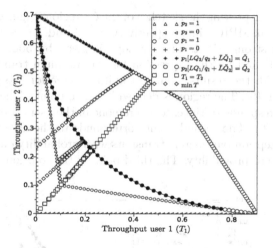

Fig. 5. Throughput region (strong MPR).

top right corner of the figure is given by the point where the user with best reception statistics transmits with probability one while the other is idle (transmission probability zero). We recall that a value of Gini-index $F_G = 1$ indicates the worst fairness situation where one of the users occupies all transmission resources. On the other hand, a value of Gini-index $F_G = 0$ indicates the maximum fairness between the users who achieve identical statistical performance. The point with the best fairness and maximum sum-throughput at the top left corner of the figure is given by the case where both users experience the same throughput. The left side boundary of the region, given by the best fairness indicator, corresponds to the curve with equal throughput in $(T_1 = T_2)$. The case

of weak MPR in Figs. 6 and 7 show that an improvement of fairness is translated into a reduction of the sum-throughput, which is consequence of the non-
convexity of the throughput region. When using CSMA (Fig. 7) we can observe
that this transition is smoother than the ALOHA case (Fig. 6), which means
that CSMA helps in achieving a better trade-off between throughput and fairness. The strong MPR case in Fig. 8 shows that higher levels of sum-throughput
can be achieved with very good levels of fairness, which is consequence of the
convexity of the throughput region. This means that strong MPR capabilities
help in achieving simultaneously higher levels of sum-throughput and improved
fairness between the users without the need of carrier sensing. The overall maximum sum-throughput in the case of strong MPR in Fig. 8 is achieved when
both users transmit simultaneously, in contrast to the weak MPR case where
the maximum sum-throughput was achieved when the user with best reception
parameters is the only active transmitter.

6.3   Sum-throughput vs. Power Region

Figures 9, 10 and 11 present the sketches of the C_P region. Figures 9 and 10 show
the results for weak MPR using, respectively, $L = 1$ and $L = 8$. Figure 11 shows
the results with strong MPR. The envelope is given by boundary conditions
$((p_1, p_2) \in \{0, 1\})$ and by the expression for the optimum transmission probabilities in (23). All the results assume a unitary value for the proportionality
constant $\alpha = 1$ in (9). The region is defined by three points. The first one is the
origin which corresponds to the case where none of the users transmits information $(p_1 = p_2 = 0)$. The second point corresponds to the case where the user
with the best reception parameters transmits with probability one and the other
transmits with zero probability. The third point is the solution with maximum

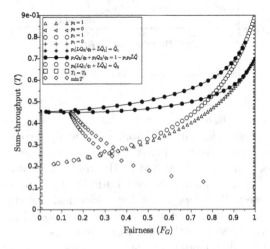

Fig. 6. Sum-throughput vs. fairness (weak MPR) with $L = 1$ (ALOHA).

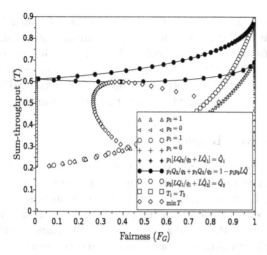

Fig. 7. Sum-throughput vs. fairness (weak MPR) with $L = 8$ (CSMA).

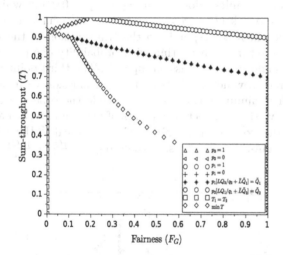

Fig. 8. Sum-throughput vs. fairness (strong MPR).

power consumption ($P = 2$), which is given by the case where both users transmit with probability one ($p_1 = p_2 = 1$). The bottom boundary of the region is defined by the curve in (23). The line that defines the top left boundary of the region is defined by the case where the user with the worst reception parameters remains silent (zero transmission probability), while the other user increases its transmission probability. The line that defines the right top boundary is given by the case where the user with the lowest reception parameters starts increasing its transmission probability while the other user keeps transmitting with probability equal to one. This leads either to a gradual reduction (in the case of weak

MPR in Figs. 9 and 10) or a gradual increase (in case of strong MPR in Fig. 11) of sum-throughput until the line reaches the maximum power consumption value $P = 2$ at the right hand side of the region. Note that in the case of weak MPR in Figs. 9 and 10, the point of maximum power transmission yields a relatively low value of sum-throughput. By contrast, in the case of strong MPR this situation is changed, as maximum power yields also maximum sum-throughput.

6.4   Return vs. Risk Region

Figures 12, 13 and 14 present the sketches of the return vs. risk trade-off region whose envelope is given in parametric form by the expressions for return in (11), risk in (12), and the expression for the optimum transmission probabilities in (26). The curves were obtained by using the following economic values: $\hat{r}_1 = 0.4$, $E[(r_1 - \hat{r}_1)^2] = 0.1$, $\hat{r}_2 = 0.55$, and $E[(r_2 - \hat{r}_2)^2] = 0.5$. These values correspond to a scenario where the transmissions with highest return are also the transmissions with the highest financial risk. The region in the weak MPR case in Figs. 12 and 13 is defined by three points. The origin, which corresponds to the case where none of the users transmits information ($p_1 = p_2 = 0$), and which yields zero return and also zero risk ($R = S = 0$). The second point is the solution with maximum return which corresponds to the transmission of the user with the maximum average return transmitting with probability one ($p_2 = 1$) while the remaining user is idle (no transmission, $p_1 = 0$). The third point is the solution with minimum return which is given by the transmission with probability one of the user with minimum return ($p_2 = 1$) while the other user remains idle ($p_1 = 0$). The curve that connects the points of maximum return and maximum risk is given by the expression in (26) which is identical to the solution for the throughput and sum-throughput vs. fairness region $\frac{p_1 Q_2}{q_2} + \frac{p_2 Q_1}{q_1} = 1 - \bar{p}_1 \bar{p}_2 \bar{Q}$.

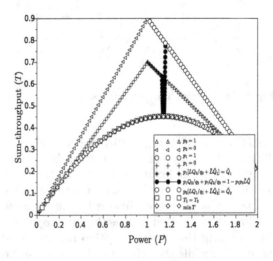

Fig. 9. Sum-throughput vs. power (weak MPR) with $L = 1$ (ALOHA).

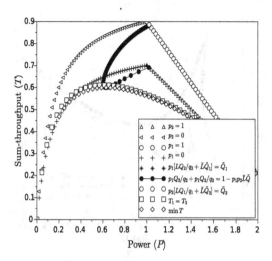

Fig. 10. Sum-throughput vs. power (weak MPR) with $L = 8$ (CSMA).

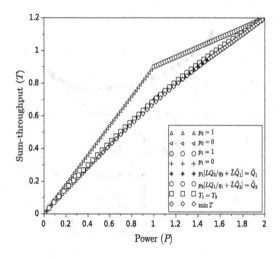

Fig. 11. Sum-throughput vs. power (strong MPR).

We can conclude then that for ALOHA systems with two users, the Pareto optimal trade-off front of the return vs. risk region is projected exactly over the Pareto optimal trade-off front of the throughput region. This does not mean, however, that the trade-off fronts are identical. It can be observed that the transmissions of the user with the higher risk make the trade-off front to be enlarged towards the right side of the figure, which means that risk grows faster and thus in terms of trade-off it becomes different to the trade-off provided by the envelope of the throughput region. The curve that connects the origin with the point of maximum return can be obtained by keeping the user with lower

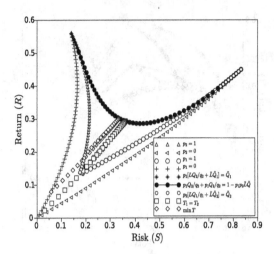

Fig. 12. Return vs. risk (weak MPR) with $L = 1$ (ALOHA).

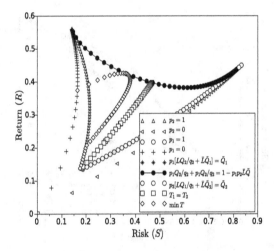

Fig. 13. Return vs. risk (weak MPR) with $L = 8$ (CSMA).

return with zero transmission probability ($p_2 = 0$) and the user with maximum return with variable transmission probability. Similarly, the curve that connects the origin with the point of minimum return is obtained by keeping the user with minimum return with zero transmission probability ($p_1 = 0$) and the user with maximum risk with variable transmission probability. The effects of carrier sense in the weak MPR case can be observed in Fig. 13 where the Pareto trade-off curve provides a better trade-off between return and risk than in Fig. 12. Note that it is possible to obtain higher values of return with lower values of risk. The strong MPR case displayed in Fig. 14 shows that the convex throughput region

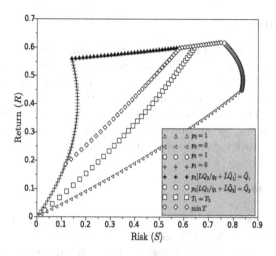

Fig. 14. Return vs. risk (strong MPR).

is translated into a considerable increase of the area of the return vs. risk region, which means that higher values of return can be achieved with considerable lower values of risk. The region is now defined by the three points described for the weak MPR case plus the maximum point of sum-throughput which is achieved when both users transmit and can be decoded simultaneously most of the time thanks to the MPR capabilities of the system.

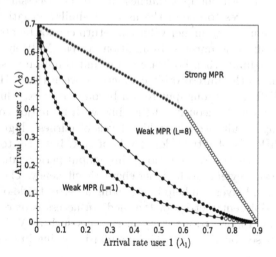

Fig. 15. Stability region

6.5   Stability Region

The stability region displayed in Fig. 15 shows the similarity with the throughput region displayed in Fig. 3, which is a result expected from previous analysis in the literature. The non-convexity of the stability region indicates a strong contention when the users collide with each other or when they experience simultaneously high traffic loads. The optimization method and the analysis provided in the previous subsection for the stability region can be used to obtain the envelope of the stability region for systems with more than two users too.

7   Conclusions

This paper has presented a trade-off analysis of different types of metrics of a CSMA protocol with MPR using multi-objective and financial portfolio optimization tools. The trade-off regions investigated were: the conventional throughput and stability regions, sum-throughput vs. fairness, sum-throughput vs. power, and return vs. risk. The throughput region was found to be non-convex in the case of weak MPR and convex in the case of strong MPR. Carrier sensing was found to reduce the non-convexity of the region in the weak MPR case. The projection of the curve that describes the top boundary of the throughput region was found to also describe the top boundary of the fairness region and the return-risk region. The non-convexity of the weak MPR case causes the system not to achieve simultaneously good levels of fairness and maximum sum-throughput, or higher levels of return with low levels of risk. This means that when we desire to increase sum-throughput, fairness has to be necessarily sacrificed by providing more resources to one of the users. A similar situation arises in the return vs. risk region, where higher values of return cannot be obtained without increasing risk or affecting fairness. In addition, a weak MPR system shows low levels of sum-throughput when both users transmit simultaneously (high power consumption) due to the unresolvable conflicts between them. However, in the case of strong MPR, the throughput region becomes convex, which means that users can transmit simultaneously all the time with no major conflict between them, thus leading simultaneously to high levels of fairness, high levels of sum-throughput, high levels of return, low levels of risk, but also to high levels of power consumption. However, the ratio of throughput performance to consumed power can be shown to yield much higher energetic efficiency. These results show that a strong physical layer can help in achieving a better trade-off in the Pareto optimal sense of different metrics of the medium access control (MAC) layer. The results can be applied to a variety of systems such as WiFi, and wireless sensor networks based on IEEE 802.15.4 standard working in contention mode.

Acknowledgments. The research leading to these results has received funding from the ARTEMIS Joint Undertaking under grant agreement no. 621353, the Portuguese National Science Foundation FCT, and by the North Portugal Regional Operational Programme (ON.2 O Novo Norte), under the National Strategic Reference Framework

(NSRF), through the European Regional Development Fund (ERDF), and by FCT, within project ref. NORTE-07-0124-FEDER-000063 (BEST-CASE, New Frontiers).

References

1. Deliverable D6.5.1: Specification of cognitive and opportunistic functions of the spectrum management framework, FP7 QoSMOS: Quality of Service and MObility driven cognitive radio Systems. http://www.ict-qosmos.eu
2. Karla, I.: Resolving SON interactions via self-learning prediction in cellular wireless networks. In: WICOM, Shangai (2012)
3. Abramson N.: The ALOHA system - another alternative for computer communications. In: Proceedings of the 1970 Fall Joint Computer Conference. AFIPS Press (1970)
4. Naware, V., Mergen, G., Tong, L.: Stability and delay of finite-user slotted ALOHA with multipacket reception. IEEE Trans. Inf. Theory $51(7)$, 2636–2656 (2005)
5. Samano-Robles, R., Gameiro, A.: A Slotted-ALOHA protocol with cooperative diversity. In: 4th Annual Wireless Internet Conference WICON 2008, Maui, Hawai, 21 (2008)
6. Baccelli, F.: Stochastic analysis of spatial and opportunistic aloha. IEEE J. Sel. Areas Commun. $27(7)$, 1105–1119 (2009)
7. Tobagi, F.A., Kleinrock, L.: Packet switching in radio channels: part IV-stability considerations and dynamic control in carrier sense multiple access. IEEE Trans. Commun. $25(10)$, 1103–1119 (1977)
8. Samano-Robles, R., Gameiro, A.: The throughput region of wireless random access protocols with multipacket reception. In: Proceedings of the International Workshop on Telecommunications, Sao Paulo, Brazil, vol. 1, pp. 207–212 (2009)
9. Samano-Robles, R., Ghogho, M., McLernon, D.C.: Wireless networks with retransmission diversity and carrier sense multiple access. IEEE Trans. Sig. Proc. $57(9)$, 3722–3726 (2009)
10. Kleinrock, L., Tobagi, F.A.: Packet switching in radio channels: Part I : carrier sense multiple access modes and their throughput-delay characteristics. IEEE Trans. Commun. $23(12)$, 1400–1416 (1975)
11. Bruno, R., Conti, M., Gregori, E.: Optimization of efficiency and energy consumption in p-persistent CSMA-based wireless LANs. IEEE Trans. Mob. Comput. $1(1)$, 10–31 (2002)
12. Marshall, A.W., Olkin, I.: Inequalities: Theory of Majorization and Its Applications. Academic Press, New York (1979)
13. Sen, A.: On Economic Inequality. Clarendon Press, Oxford (1973)
14. Elton, E.J., Gruber, M.J., Brown, S.J., Goetzmann, W.N.: Modern Portfolio Theory and Investment Analysis. Wiley, Hoboken (2004)
15. Roberts, L.G.: ALOHA packet system with and without slots and capture. Comput. Commun. Rev. $5(2)$, 28–42 (1975)
16. Rao, R., Ephremides, A.: On the stability of interacting queues in a multiple-access system. IEEE Trans. Inf. Theory $4(5)$, 918–930 (1988)
17. Zorzi, M., Rao, R.: Capture and retransmission control in mobile radio. IEEE J. Sel. Areas Commun. $12(8)$, 1289–1298 (1994)
18. Yu, Y., Cai, X., Giannakis, G.B.: On the stability of slotted ALOHA with capture. IEEE Trans. Wirel. Comm. $5(2)$, 257–261 (2006)
19. Ghez, S., Verdu, S., Schwartz, S.: Stability properties of slotted Aloha with multipacket reception capability. IEEE Trans. Autom. Control $33(7)$, 640–649 (1988)

20. Ghez, S., Verdu, S., Schwartz, S.: Optimal decentralized control in the random access multipacket channel. IEEE Trans. Autom. Control **34**(11), 1153–1163 (1989)
21. Zhao, Q., Tong, L.: A dynamic queue protocol for multiaccess wireless networks with multipacket reception. IEEE Trans. Wirel. Commun. **3**(6), 2221–2231 (2004)
22. Zhao, Q., Tong, L.: A multiqueue service room MAC protocol for wireless networks with multipacket reception. IEEE Trans. Network. **11**(1), 125–137 (2003)
23. Ngo, M.H., Krishnamurthy, V., Tong, L.: Optimal channel-aware ALOHA protocol for Random Access in WLANs with multipacket reception and decentralized channel state information. IEEE Trans. Sig. Proc. **56**(6), 2575–2588 (2008)
24. Adireddy, S., Tong, L.: Exploiting decentralized channel state information for random access. IEEE Trans. Inform. Theory **51**(2), 537–561 (2005)
25. Ngo, M.H., Krishnamurty, V.: Game theoretic cross-layer transmission policies in multipacket reception wireless networks. IEEE Trans. Sig. Proc. **55**(5), 1911–1926 (2007)
26. Luo, J., Ephremides, A.: On the throughput, capacity, and stability regions of random multiple access. IEEE Trans. Info. Theory **52**(6), 2593–2607 (2006)
27. Luo, W., Ephremides, A.: Stability of N interacting queues in random-access systems. IEEE Trans. Inf. Theory **45**(5), 1579–1587 (1999)
28. Gai, Y., Ganesan, S., Krishnamachari, B.: The saturation throughput region of p-persistent CSMA. In: Information Theory and Applications Workshop (ITA), pp. 1–4, February 2011
29. Smura T.: Techno-economic modelling of wireless network and industry architectures, Doctoral dissertation, Aalto University (2012)
30. Niyato, D., Hossain, E.: Spectrum trading in cognitive radio networks: a market-equilibrium-based approach. IEEE Wirel. Commun. **15**(6), 71–80 (2008)
31. Southwell, R., Chen, X., Huang, J.: Quality of service satisfaction games for spectrum sharing. In: IEEE INFOCOM Mini Conference, Turin, Italy, pp. 570–574, April 2013
32. Chen, X., Huang, J.: Spatial spectrum access game: nash equilibria and distributed learning. In: ACM Mobihoc Hilton Head Island, South Carolina, pp. 205–214 (2012)
33. Duan, L., Huang, J., Shou, B.: Duopoly competition in dynamic spectrum leasing and pricing. IEEE Trans. Mob. Comput. **11**(11), 1706–1719 (2012)
34. Tekin, C., et al.: Atomic congestion games on graphs and their applications in networking. IEEE Trans. Network. **20**(5), 1541–1552 (2012)
35. Duan, L., Huang, J., Shou, B.: Investment and pricing with spectrum uncertainty: a cognitive operators perspective. IEEE Trans. Mob. Comput. **10**(11), 1590–1604 (2011)
36. Zhang, Y., Niyato, D., Wang, P., Hossain, E.: Auction-based resource allocation in cognitive radio systems. IEEE Commun. Mag. **50**(11), 108–120 (2012)
37. Huang, J., Berry, R., Honig, M.L.: Auction-based spectrum sharing. Springer J. Mob. Netw. Appl. **11**(3), 405–408 (2006)
38. Wysocki, T.A., Jamalipour, A.: An economic welfare preserving framework for spot pricing and hedging of spectrum rights for cognitive radio. IEEE Trans. Netw. Serv. Manage. **9**(1), 87–99 (2012)
39. Boyd, S., Vandenberghe, L.: Convex Optimization. Cambridge University Press, Cambridge (2004)

Robust Optimal Control of Dynamically Decoupled Systems via Distributed Feedbacks

Natalia Dmitruk[(✉)]

Belarusian State University, Nezavisimosti avenue 4, 220030 Minsk, Belarus
dmitrukn@bsu.by

Abstract. We consider an optimal control problem for a large-scale dynamical system represented by a team of objects with linear time-varying decoupled dynamics subject to disturbances and coupling constraints. It is assumed that centralized control is impossible and a delay in the communication network between systems is present. An algorithm for distributed feedback control is proposed. The algorithm breaks the large scale optimal control problem into sub-problems optimizing only for the inputs of the associated system. Feasibility and suboptimality of distributed control for the overall system is established and relevant data to be exchanged between the systems is analyzed.

Keywords: Optimal control · Large-scale system · Multi-agent system · Distributed feedback · Uncertainty · Algorithm

1 Introduction

Control problems for interacting dynamical systems has received a significant attention over the recent years. This is motivated by permanent progress of control techniques and computing power that allow to tackle complex large-scale problems. In various applications centralized control of such systems is impractical or impossible due to, e.g., communication restrictions. Besides specific properties of the network are not adequately addressed by a general centralized control algorithm. In these cases distributed control techniques are employed.

Many approaches have been proposed for control of linear and nonlinear systems with coupled or decoupled dynamics within distributed model predictive control (DMPC) framework (see, e.g., [1] and the references therein). In particular, in [2] for a class of discrete-time systems with coupled linear time-invariant dynamics sufficient conditions for stability of the closed-loop using stability constraints and assuming one-step communication delay are given. In [3] a distributed control strategy is obtained by solving local min-max optimization problems that treat states of the neighboring systems as disturbances and therefore minimize the worst-case local performance. In [4] an iterative cooperating distributed algorithm for linear discrete-time systems interconnected by their inputs is presented that is equivalent to the centralized controller at the limit of iterations.

© Springer International Publishing Switzerland 2015
A. Plakhov et al. (Eds.): EmC-ONS 2014, CCIS 499, pp. 95–106, 2015.
DOI: 10.1007/978-3-319-20352-2_6

Other than stabilization cooperative tasks such as consensus and synchronization are handled, e.g., by a general DMPC framework reported in [5].

Most DMPC schemes, when defining the local optimal control problems for each system, do not take into account disturbances acting on the dynamical systems. The notable exception is [6] where linear time-invariant systems subject to coupling constraints and bounded disturbances are considered and a robust DMPC scheme is proposed that implies sequential solution of local optimal control problems at each step.

In this paper we consider an optimal control problem for continuous time systems with decoupled linear time-varying dynamics subject to unknown but bounded disturbances. The systems are coupled by state constraints. The control objective on a finite control interval is to minimize the worst-case value of a given terminal penalty, though, as will be shown below, other types of performance index can be handled within the proposed approach. The idea is to incorporate distributed feedback control design into the classical optimal control problem, obtaining suboptimality of some degree and guaranteeing satisfaction of the hard constraints at each time instant. The approach presented here follows the ideas of [7,8], where dynamically coupled systems are considered. In contrast to the latter here we are able to prove recursive feasibility and suboptimality of distributed inputs.

The overall paper is structured as follows. In Sect. 2 we outline the mathematical problem formulation and the control objective for a set of linear time-varying systems subject to coupling constraints and unknown but bounded disturbances. Section 3 reviews centralized solution to this problem that guarantees robust constraint satisfaction and minimizes the worst-case performance for all possible disturbances. Section 4 presents an algorithm for robust distributed control. Feasibility and suboptimality of the distributed inputs with respect to the overall system behavior as well as communication data and requirements for the distributed algorithm are analyzed. The effectiveness of the proposed scheme is demonstrated in Sect. 5 with an illustrative example comparing performance of centralized and distributed controls. Section 6 provides some conclusions.

2   Problem Formulation

We consider a team of q continuous-time linear time-varying systems with decoupled dynamics

$$\dot{x}_i = A_i(t)x_i + B_i(t)u_i + M_i(t)w_i, \ x_i(t_0) = x_{i0}, \ t \in [t_0, t_f], \quad (1)$$

where $x_i = x_i(t) \in \mathbb{R}^{n_i}$ denotes the state of the i-th system at time t, $u_i = u_i(t) \in U_i \subset \mathbb{R}^{r_i}$ denotes the bounded control input to system i and $w_i = w_i(t) \in W_i \subset \mathbb{R}^{p_i}$ is the unknown piecewise continuous disturbance acting upon system i, $A_i(t) \in \mathbb{R}^{n_i \times n_i}$, $B_i(t) \in \mathbb{R}^{n_i \times r_i}$, $M_i(t) \in \mathbb{R}^{n_i \times p_i}$, $t \in [t_0, t_f]$, are piecewise continuous matrix functions, $i \in I = \{1, 2, \ldots, q\}$. The input constraint set U_i and the disturbance set W_i are given convex polytopes containing the origin and independent across the systems.

The input u_i is a sampled-data control that changes its value at fixed sampling instants and is constant in between. Sampling instants are in the following denoted by τ, where $\tau \in T_h = \{t_0 + kh, k = \overline{0, N-1}\}$. Here h denotes the constant sampling time defined in terms of the discretization $N \in \mathbb{N}$ of the finite control interval $[t_0, t_f]$: $h = (t_f - t_0)/N$. Thus, the input u_i in (1) is given by:

$$u_i(t) \equiv u_i(\tau), \quad t \in [\tau, \tau + h[, \ \tau \in T_h,$$

where $u_i(\tau)$ depends on the current state of system i and some exchanged information from other systems.

At time instants $s \in T_c \subseteq T_h \cup t_f$ the team is subject to coupling state constraints

$$\sum_{k \in K^l} H_k^l(s) x_k(s) \leq \alpha^l(s), \ l \in L = \{1, \ldots, l^*\}, \tag{2}$$

where $K^l \subseteq I$, $|K^l| \geq 2$; $H_k^l(s) \in \mathbb{R}^{m^l \times n_k}$, $H_k^l(s) \neq 0$ for all $k \in K^l$; $\alpha^l(s) \in \mathbb{R}^{m^l}$.

The control objective is to minimize the worst-case value of a linear terminal penalty

$$\max_{w_k, k \in I} \sum_{k \in I} c_k^T x_k(t_f), \tag{3}$$

while satisfying the decoupled input and coupling state constraints (2).

In the following we have to distinguish between the variables used in the optimal control problems for predictions and the real system/plant variables. To this end the latter will be denoted by a superscript $*$. Thus, u_i^* and x_i^* denote the input and the state trajectory which realize in a particular control process, and w_i^* denotes a realized unknown disturbance. It is assumed that at all time instants $\tau \in T_h \cup t_f$ the current state $x_i^*(\tau)$ is completely measured by system i.

3   Centralized Optimal Control

In this section we review some results from [9] on centralized optimal feedback control of dynamical systems subject to bounded disturbances that are needed in the later sections.

When centralized control is implemented, one central controller chooses the inputs for all systems (1) in the team, treating the problem under consideration as a large-scale optimal control problem without taking into account its decoupled dynamics or a specific interconnection structure. The overall system dynamics is then represented in concatenated form

$$\dot{x} = A(t)x + B(t)u + M(t)w, \quad x(t_0) = x_0, \tag{4}$$

where $x(t) \in \mathbb{R}^n$, $u(t) \in \mathbb{R}^r$ and $w(t) \in \mathbb{R}^p$ with $n = \sum_{k \in I} n_k$, $r = \sum_{k \in I} r_k$, $p = \sum_{k \in I} p_k$, denote the state, the input and the disturbance of the overall system at time t, i.e. $x(t) = (x_1(t), \ldots, x_q(t))$, $u(t) = (u_1(t), \ldots, u_q(t))$, $w(t) =$

$(w_1(t), \ldots, w_q(t))$; $A(t) \in \mathbb{R}^{n \times n}$, $B(t) \in \mathbb{R}^{n \times r}$, $M(t) \in \mathbb{R}^{n \times p}$, $t \in [t_0, t_f]$, are the corresponding block diagonal matrices.

In this paper, both in centralized and distributed control schemes, we use only one type of feedback that can be defined for uncertain systems (see e.g. [9, 10]), namely the open-loop optimal feedback, which refers to the fact that a feedback strategy is obtained via repetitive (for every time instant $\tau \in T_h$) solution of an open-loop min-max optimal control problem subject to a shrinking control interval $[\tau, t_f]$ and a current overall state $x^*(\tau)$.

The open-loop min-max optimal control problem (centralized) that is solved at time τ is denoted by $\mathcal{P}(\tau)$ and has the form

$$\mathcal{P}(\tau): \qquad J^0(\tau) = \min_u \max_w c^T x(t_f), \qquad (5)$$

subject to

$$\dot{x} = A(t)x + B(t)u + M(t)w, \quad x(\tau) = x^*(\tau),$$

$$H(s)x(s) \leq \alpha(s), \quad s \in T_c(\tau) = T_c \cap [\tau, t_f], \quad u(t) \in U, \quad w(t) \in W, \quad t \in [\tau, t_f],$$

where $c = (c_1, \ldots, c_q)$; $H(s) = \begin{pmatrix} H_k^l(s), k \in I \\ l \in L \end{pmatrix} \in \mathbb{R}^{m \times n}$, $m = \sum_{l \in L} m^l$, with $H_k^l(s)$ being zero for $k \notin K^l$; $\alpha(s) = (\alpha^l(s), l \in L)$; $U = U_1 \times \ldots \times U_q$, $W = W_1 \times \ldots \times W_q$.

The optimal open-loop control of $\mathcal{P}(\tau)$ is an input $u^0(t|\tau)$, $t \in [\tau, t_f]$, such that for every realization of the disturbance $w(t) \in W$, $t \in [\tau, t_f]$, the state constraints are satisfied and the worst-case cost is minimized.

Assumption 1. Problem (5) is feasible for $\tau = t_0$.

Under Assumption 1 problem $\mathcal{P}(\tau)$ is feasible for all $\tau \in T_h$ and the centralized optimal feedback control algorithm is specified as follows [9]:

Algorithm 1. (*centralized*)

(1) Set $\tau = t_0$, $x^*(\tau) = x_0$.
(2) Find a solution $u^0(t|\tau)$, $t \in [\tau, t_f]$, to the centralized problem $P(\tau)$.
(3) Apply input $u^*(t) \equiv u^*(\tau) = u^0(\tau|\tau)$, $t \in [\tau, \tau + h[$, to the overall system.
(4) Set $\tau := \tau + h$. If $\tau < t_f$ return to step 2, else stop.

Now we briefly review how the min-max problem $P(\tau)$ is solved. Following [9], problem (5) can be reduced to a deterministic optimal control problem for nominal system (i.e. system (4) without the disturbance term) which constraints are tightened to ensure robust feasibility of the inputs in (5).

Denote by $F(t) \in \mathbb{R}^{n \times n}$, $t \in [t_0, t_f]$, the fundamental matrix of the overall system (4): $\dot{F}(t) = A(t)F(t)$, $F(t_0) = I^n$, where $I^n \in \mathbb{R}^{n \times n}$ is an identity matrix.

For a given input $u(t)$, $t \in [\tau, t_f]$, and disturbance $w(t)$, $t \in [\tau, t_f]$, the overall output $y(s) = H(s)x(s)$ at time instant $s \in T_c(\tau)$, can be found as

$$y(s) = H(s)F(s)F^{-1}(\tau)x^*(\tau) + \int_\tau^s H(s)F(s)F^{-1}(t)[B(t)u(t) + M(t)w(t)]dt.$$

Introduce matrix functions $\Phi(s,t) \in R^{m \times n}$, $t \in [t_0, s]$, $k \in I$, such that $\Phi(s,t) = H(s)F(s)F^{-1}(t)$. Obviously, $\partial\Phi(s,t)/\partial t = -\Phi(s,t)A(t)$, $\Phi(s,s) = H(s)$. Then

$$y(s) = \Phi(s,\tau)x^*(\tau) + \int_\tau^s \Phi(s,t)[B(t)u(t) + M(t)w(t)]dt.$$

The input $u(t)$, $t \in [\tau, t_f]$, is feasible in (5) for every possible realization of the disturbance w, if and only if $y(s) \le \alpha(s)$ for all $s \in T_c(\tau)$ which translate into the inequalities

$$\Phi(s,\tau)x^*(\tau) + \int_\tau^s \Phi(s,t)B(t)u(t)dt + \gamma(s|\tau) \le \alpha(s), \ s \in T_c(\tau).$$

Here the first two terms are the output of the overall nominal system (4) and the term $\gamma(s|\tau) \in \mathbb{R}^m$, $s \in T_c(\tau)$, corresponds to the worst-case realization of the disturbances: $\gamma(s|\tau) = (\gamma_j(s|\tau), j = \overline{1,m})$,

$$\gamma_j(s|\tau) = \int_\tau^s \max_{w \in W} \phi_j(s,t)^T M(t)wdt,$$

where $\phi_j(s,t)^T$ is the j-th row of the matrix $\Phi(s,t)$.

Concluding, the optimal open-loop control $u^0(t|\tau)$, $t \in [\tau, t_f]$, of problem $\mathcal{P}(\tau)$ is obtained by the solution of the deterministic optimal control problem

$$\min_u c^T x(t_f), \tag{6}$$

subject to

$$\dot{x} = A(t)x + B(t)u, \ x(\tau) = x^*(\tau),$$

$$H(s)x(s) \le \alpha(s) - \gamma(s|\tau), \ s \in T_c(\tau), \ u(t) \in U, \ t \in [\tau, t_f].$$

The resulting cost of problem $\mathcal{P}(\tau)$ is given by

$$J^0(\tau) = \gamma^0(\tau) + c^T x^0(t_f|\tau) = \gamma^0(\tau) + \phi^0(\tau)^T x^*(\tau) + \int_\tau^{t_f} \phi^0(t)^T B(t)u^0(t|\tau)dt,$$

where $x^0(t|\tau)$, $t \in [\tau, t_f]$, is the optimal overall trajectory of (6) and $\gamma^0(\tau) = \int_\tau^{t_f} \max_{w \in W} \phi^0(t)^T M(t)wdt$, $\phi^0(t)^T = c^T F(t_f)F^{-1}(t)$, $t \in [t_0, t_f]$.

4 Distributed Optimal Control

In this section an algorithm for distributed optimal feedback control of a team of systems (1) is developed. Each system predicts its future control inputs on the base of its own current state $x_i^*(\tau)$ and some information received from neighboring systems, where neighbors are defined by the coupling constraints. It is assumed that there is a communication delay equal to the sampling time h. A centralized controller described in Sect. 3 is employed offline at initialization stage and is not available for any online computations.

4.1 Local Optimal Control Problem

To achieve control of systems (1) in a distributed fashion we associate an optimal control problem $\mathcal{P}_i(\tau)$ with each system i, minimizing over only local inputs u_i subject to local and coupling constraints. To formulate such a local optimal control problem we first define the interconnection topology for the multi-agent system under consideration.

Systems i and j are coupled by the constraints and are called neighbors if they enter the same constraint in (2). Denote by L_i all indices of the constraints (2) containing a term for system i, i.e. $L_i = \{l \in L : i \in K^l\}$. Then $N_i = \cup_{l \in L_i} K^l \setminus \{i\}$ is a set of indices of all neighbors of system i. Note that the interconnection topology here is time-invariant. It is assumed that system i can communicate only to its neighbors $k \in N_i$. The information that is exchanged over the communication network will be specified in Sect. 4.2.

In the following the $u_i^d(\cdot|\tau) = (u_i^d(t|\tau),\, t \in [\tau, t_f])$ denotes the distributed input predicted by system i at time τ, i.e. the optimal open-loop control of local problem $\mathcal{P}_i(\tau)$. Concatenated distributed input $u^d(\cdot|\tau) = (u_k^d(\cdot|\tau), k \in I)$ will be also referred to as the optimal distributed open-loop control. The corresponding state trajectory of the nominal system (1) with the initial state $x_i(\tau) = x_i^*(\tau)$ is denoted by $x_i^d(\cdot|\tau) = (x_i^d(t|\tau), t \in [\tau, t_f])$. Furthermore, $y_i^l(s|\tau) = H_i^l(s)x_i^d(s|\tau)$, $s \in T_c(\tau),\, l \in L_i$, denote the outputs of system i predicted at time τ. The overall distributed output corresponding to the l-th constraint (2) at time instant $s \in T_c$ is $y^l(s|\tau) = \sum_{k \in K^l} y_k^l(s|\tau)$.

Following [8], define the open-loop min-max optimal control problem $\mathcal{P}_i(\tau)$ for system i at time instant $\tau \in T_h \setminus t_0$:

$$\mathcal{P}_i(\tau): \qquad J_i(\tau) = \min_{u_i} \max_{w_i} \sum_{k \in I} c_k^T x_k(t_f),$$

subject to

$$\dot{x}_i = A_i(t)x_i + B_i(t)u_i + M_i(t)w_i,\ x_i(\tau) = x_i^*(\tau),$$

$$\dot{x}_k = A_k(t)x_k + B_k(t)u_k^d(t|\tau - h),\ x_k(\tau) = x_k^d(\tau|\tau - h),\ k \in N_i,$$

$$\sum_{k \in K^l} H_k^l(s)x_k(s) \le \alpha_i^l(s|\tau),\ s \in T_c(\tau), l \in L_i,$$

$$u_i(t) \in U_i,\ w_i(t) \in W_i,\ t \in [\tau, t_f]. \tag{7}$$

Here the input u_i of the i-th system is the optimization variable, and the inputs u_k of systems $k \in I_i$ are held as fixed parameters equal to their distributed inputs $u_k^d(\cdot|\tau - h)$ predicted at the previous time $\tau - h$. Thus, system i assumes that its neighbors keep controls predicted at time $\tau - h$ also for the current time τ and besides they follow nominal trajectories, i.e. $w_k(t) \equiv 0,\ t \in [\tau - h, t_f]$, $k \in N_i$. Then their predicted states $x_k^d(\tau|\tau - h)$ are used in $\mathcal{P}_i(\tau)$ as initial states at time instant τ. The initial state of system i is its current state $x_i^*(\tau)$.

In $\mathcal{P}_i(\tau)$ the coupling state constraints have a modified right hand side

$$\alpha_i^l(s|\tau) = y^l(s|\tau - h) + \Omega_i^l(s|\tau)[\alpha^l - y^l(s|\tau - h)],\ l \in L_i, \tag{8}$$

where $\Omega_i^l(\tau) \in \mathbb{R}^{m^l \times m^l}$ is a diagonal matrix of weight parameters for system i in constraint l, $\sum_{k \in K^l} \Omega_k^l(s|\tau) = I^{m^l}$ for all $s \in T_c(\tau)$, $l \in L$.

The idea behind constraints modifications (8) is to guarantee feasibility of the optimal distributed open-loop control $u^d(\cdot|\tau)$ with respect to the overall system. This feasibility result is proved in Sect. 4.3.

In can be seen from (7) and (8) that in order to construct problem $\mathcal{P}_i(\tau)$ system i needs to know the dynamics of the neighboring systems, their predicted states $x_k^d(\tau|\tau - h)$, whole input trajectories $u_k^d(\cdot|\tau - h)$, and the outputs $y_k^l(s|\tau - h)$, $s \in T_c(\tau)$, $k \in N_i$. However, in contrast to [8], where systems with coupled dynamics are studied, some information here is abundant. In the next section we derive an equivalent formulation of problem $\mathcal{P}_i(\tau)$ that compared to (7) has a reduced dimension and requires less data from other systems.

4.2   An Equivalent Formulation of $\mathcal{P}_i(\tau)$ and the Algorithm

Since in problem $\mathcal{P}_i(\tau)$ dynamics of systems $k \in N_i$ is deterministic and doesn't depend on input u_i, their trajectories $x_k^d(\cdot|\tau - h)$ are known parameters. They can be excluded from the dynamics (7) and embedded into the modified state constraints. The latter take the form

$$H_i^l(s)x_i(s) + \sum_{k \in K^l \setminus i} H_k^l(s)x_k^d(s|\tau - h) =$$

$$= H_i^l(s)x_i(s) + \sum_{k \in K^l \setminus i} y_k^l(s|\tau - h) \le \alpha_i^l(\tau), \ s \in T_c(\tau), \ l \in L_i.$$

Denote $\bar{\alpha}_i^l(\tau) = \alpha_i^l(\tau) - \sum_{k \in K^l \setminus i} y_k^l(s|\tau - h) = y_i^l(s|\tau - h) + \Omega_i^l(s|\tau)(\alpha^l(s) - y^l(s|\tau - h))$ to obtain the new state constraints

$$H_i^l(s)x_i(s) \le \bar{\alpha}_i^l(s|\tau), \ s \in T_c(\tau), \ l \in L_i.$$

The resulting local optimal control problem for system i at time $\tau \in T_h \setminus t_0$ is

$$\mathcal{P}_i^d(\tau): \qquad J_i^d(\tau) = \min_{u_i} \max_{w_i} c_i^T x_i(t_f),$$

subject to

$$\dot{x}_i = A_i(t)x_i + B_i(t)u_i + M_i(t)w_i, \ x_i(\tau) = x_i^*(\tau),$$

$$H_i^l(s)x_i(s) \le \bar{\alpha}_i^l(s|\tau), \ s \in T_c(\tau), \ l \in L_i,$$

$$u_i(t) \in U_i, \ w_i(t) \in W_i, \ t \in [\tau, t_f].$$

Similarly to Sect. 3, problem $\mathcal{P}_i^d(\tau)$ can be reduced to a deterministic problem for the nominal system (1) with the tightened constraints:

$$\min_{u_i} c_i^T x_i(t_f),$$

subject to

$$\dot{x}_i = A_i(t)x_i + B_i(t)u_i, \ x_i(\tau) = x_i^*(\tau),$$

$$H_i^l(s)x_i(s) \le \bar{\alpha}_i^l(s|\tau) - \gamma_i^l(s|\tau), \ s \in T_c(\tau), \ l \in L_i, \ u_i(t) \in U_i, \ t \in [\tau, t_f].$$

where $\gamma_i^l(s|\tau) = (\gamma_{i,j}^l(s|\tau), j = \overline{1,m^l})$: $\gamma_{i,j}^l(s|\tau) = \int_\tau^s \max_{w_i \in W_i} \phi_{i,j}^l(s,t)^T M_i(t)w_i dt$,

and $\phi_{i,j}^l(s,t)^T$ is the j-th row of the matrix $\Phi_i^l(s,t) = H_i^l(s)F_i(s)F_i^{-1}(t)$, and $F_i(t) \in \mathbb{R}^{n_i \times n_i}$, $t \in [t_0, t_f]$, denotes the fundamental matrix of system (1): $\dot{F}_i(t) = A_i(t)F_i(t), F_i(t_0) = I^{n_i}$.

The cost of problem $\mathcal{P}_i^d(\tau)$ is equal to

$$J_i^d(\tau) = \gamma_i^0(\tau) + \phi_i^0(\tau)^T x_i^*(\tau) + \int_\tau^{t_f} \phi_i^0(t)^T B_i(t)u_i^d(t|\tau)dt, \qquad (9)$$

where $\gamma_i^0(\tau) = \int_\tau^{t_f} \max_{w_i \in W_i} \phi_i^0(t)^T M_i(t)w_i dt$, $\phi_i^0(t)^T = c_i^T F_i(t_f)F_i^{-1}(t)$.

Analyzing problem $\mathcal{P}_i^d(\tau)$ we conclude that system i needs the following information at time $\tau \in T_h \setminus t_0$:

(1) its complete current state $x_i^*(\tau)$;
(2) from all neighboring systems $k \in N_i$ delayed by h
the outputs $y_k^l(s|\tau - h) = H_k^l(s)x_k^d(s|\tau - h)$, $l \in L_i$, corresponding to the distributed input $u_k^d(\cdot|\tau - h)$ predicted at time $\tau - h$.

The distributed optimal feedback control algorithm is specified as follows:

Algorithm 2. (*distributed*)

(1) Set $\tau = t_0$, $x^*(\tau) = x_0$.
(2) Find a solution $u^0(\cdot|t_0)$ to the centralized problem $P(t_0)$ and set $u_i^d(t|t_0) = u_i^0(t|t_0)$, $t \in [t_0, t_f]$, $i \in I$.
For each system $i \in I$ (in parallel):
(3) Apply input $u_i^*(t) \equiv u_i^*(\tau) = u_i^d(\tau|\tau)$, $t \in [\tau, \tau + h[$.
(4) Communicate the outputs $y_i^l(\tau)$ to neighbors $k \in K^l \setminus i$, $l \in L_i$.
(5) Set $\tau := \tau + h$. If $\tau = t_f$ stop.
(6) Solve problem $\mathcal{P}_i^d(\tau)$ to find $u_i^d(t|\tau)$, $t \in [\tau, t_f]$. Return to step 3.

4.3 Properties of Distributed Control

The important properties that a distributed control scheme should possess are feasibility of distributed inputs with respect to coupling constraints (2), suboptimality of the distributed input with respect to a centralized one, and a recursive feasibility of the optimal control problems $\mathcal{P}_i^d(\tau)$ solved at each sampling time $\tau \in T_h \setminus t_0$. In this section we prove these properties for Algorithm 2.

Theorem 1. *For any* $\tau \in T_h \setminus t_0$ *the optimal distributed open-loop control* $u^d(\cdot|\tau) = (u_k(\cdot|\tau), k \in I)$ *is a feasible input in the centralized optimal control problem* $\mathcal{P}(\tau)$.

Proof. Let instant τ be fixed. We have to prove that the overall distributed trajectory $x^d(\cdot|\tau)$, corresponding to the input $u^d(\cdot|\tau)$ satisfies the inequalities

$$H(s)x^d(s|\tau) \le \alpha - \gamma(s|\tau), s \in T_c(\tau), \qquad (10)$$

as defined by problem (6) equivalent to $\mathcal{P}(\tau)$.

Consider problem $\mathcal{P}_k^d(\tau)$. Its optimal open-loop control $u_k^d(\cdot|\tau)$ is feasible, therefore the corresponding optimal trajectory $x_k^d(\cdot|\tau)$ satisfies the inequalities

$$H_k^l(s)x_k^d(s|\tau) \leq \bar{\alpha}_k^l(s|\tau) - \gamma_k^l(s|\tau), \quad s \in T_c(\tau), \ l \in L_i. \tag{11}$$

Summing (11) over all $k \in K^l$ for a fixed $l \in L$ and taking into account that

$$\sum_{k \in K^l} \bar{\alpha}_k^l(s|\tau) = \sum_{k \in K^l} [y_k^l(s|\tau - h) + \Omega_k^l(s|\tau)(\alpha^l(s) - y^l(s|\tau - h))] = \alpha^l(s),$$

$$\gamma(s|\tau) = (\gamma^l(s|\tau), l \in L), \quad \gamma^l(s|\tau) = \sum_{i \in K^l} \gamma_i^l(s|\tau), \ l \in L,$$

obtain

$$\sum_{k \in K^l} H_k^l(s)x_k^d(s|\tau) \leq \alpha^l(s) - \gamma^l(s|\tau), \ l \in L.$$

The latter in concatenated form is given by (10), therefore $u^d(\cdot|\tau)$ is feasible in the centralized problem $\mathcal{P}(\tau)$. □

To guarantee recursive feasibility of the distributed algorithm, i.e. existence of solution of problem $\mathcal{P}_i^d(\tau)$ for all $\tau \in T_h \setminus t_0$, we assume:

Assumption 2. $\Omega_i^l(s|\tau)$ is such that $\gamma_i^l(s|\tau - h) = \Omega_i^l(s|\tau)\gamma^l(s|\tau - h)$.

The following theorem implies that if a centralized solution for $\tau = t_0$ exists and the weights are properly chosen, then Algorithm 2 can indeed be implemented for distributed feedback control.

Theorem 2. *Under Assumptions 1,2 problem $\mathcal{P}_i^d(\tau)$ is feasible for all $\tau \in T_h \setminus t_0$.*

Proof. To prove the assertion it is sufficient to show that the distributed input $u_i^d(\cdot|\tau - h)$, predicted by system i at time $\tau - h$, is feasible in problem $\mathcal{P}_i^d(\tau)$. Then according to optimal control existence theorems there exists a solution $u_i^d(\cdot|\tau)$ to problem $\mathcal{P}_i^d(\tau)$.

For the trajectory $x_i(\cdot|\tau)$ of nominal system (1) corresponding to the control $u_i^d(\cdot|\tau - h)$ the following is true

$$H_i^l(s)x_i(s|\tau) = \Phi_i^l(s,\tau)x_i^*(\tau) + \int_\tau^s \Phi_i^l(s,t)B_i(t)u_i^d(t|\tau - h)dt = y_i^l(s|\tau - h) +$$

$$+ \Phi_i^l(s,\tau)(x_i^*(\tau) - x_i^d(\tau|\tau - h)) = y_i^l(s|\tau - h) + \int_{\tau-h}^\tau \Phi_i^l(s,t)M_i(t)w_i^*(t)dt,$$

where, due to Assumption 2,

$$\int_{\tau-h}^\tau \Phi_i^l(s,t)M_i(t)w_i^*(t)dt \leq \gamma_i^l(s|\tau - h) - \gamma_i^l(s|\tau) = \Omega_i^l(s|\tau)\gamma^l(s|\tau - h) - \gamma_i^l(s|\tau).$$

Theorem 1 implies that

$$y^l(s|\tau - h) = \sum_{k \in K^l} H_k^l(s)x_k^d(s|\tau - h) \leq \alpha^l(s) - \gamma^l(s|\tau - h),$$

therefore $\gamma^l(s|\tau - h) \leq \alpha^l(s) - y^l(s|\tau - h)$.

Concluding,

$$H_i^l(s)x_i(s|\tau) \le y_i^l(s|\tau - h) + \Omega_i^l(s|\tau)\gamma^l(s|\tau - h) - \gamma_i^l(s|\tau) \le$$

$$\le y_i^l(s|\tau - h) + \Omega_i^l(s|\tau)[\alpha^l(s) - y^l(s|\tau - h)] - \gamma_i(s|\tau) = \bar{\alpha}_i^l(s|\tau) - \gamma_i^l(s|\tau).$$

Thus, $u_i^d(\cdot|\tau - h)$ satisfies the constraints of $\mathcal{P}_i^d(\tau)$ and the latter is feasible. $\quad\square$

Note that the feasibility results do not use optimality of distributed inputs and therefore hold for any type of performance index, not only for (3). Its linearity is, however, important for deriving the suboptimality properties of the distributed scheme.

Since according to Theorem 1 the optimal distributed open-loop control $u^d(\cdot|\tau)$ is a feasible input in problem $\mathcal{P}(\tau)$, one can calculate its resulting worst-case cost (3) as $\sum_{k\in I} J_k^d(\tau)$, where $J_i^d(\tau)$ is the optimal value of the performance index of problem $\mathcal{P}_i^d(\tau)$ as defined by (9).

The following theorem asserts that the cost, corresponding to the optimal distributed open-loop control $u^d(\cdot|\tau)$, is a nonincreasing function of τ.

Theorem 3. *Under Assumptions 1,2 the inequalities hold*

$$J^0(\tau) \le \sum\nolimits_{k\in I} J_k^d(\tau) \le \sum\nolimits_{k\in I} J_k^d(\tau - h) \le J^0(t_0).$$

Proof. The first inequality is a consequence of the fact, that $u^d(\cdot|\tau)$ is feasible, but not necessarily optimal in $\mathcal{P}(\tau)$.

The second inequality is obtained via the following arguments. Since, according to Theorem 2, $u_k^d(\cdot|\tau - h)$ is feasible in problem $\mathcal{P}_k^d(\tau)$, its resulting cost is not less that the optimal value $J_k^d(\tau)$:

$$J_k^d(\tau) \le \phi_k^0(\tau)^T x_k^*(\tau) + \int_\tau^{t_f} \phi_k^0(t)^T B_k(t)u_k^d(t|\tau - h) + \gamma_k^0(\tau).$$

Summing over all $k \in I$, obtain

$$\sum_{k\in I} J_k^d(\tau) \le \phi^0(\tau)^T x^*(\tau) + \int_\tau^{t_f} \phi^0(t)^T B(t)u^d(t|\tau - h) + \gamma^0(\tau)$$

$$= \phi^0(\tau)^T[x^*(\tau) - x^d(\tau|\tau - h)] + \sum_{k\in I} J_k^d(\tau - h) - \gamma^0(\tau - h) + \gamma^0(\tau) \le \sum_{k\in I} J_k^d(\tau - h).$$

The third inequality results from initialization of the distributed algorithm with the centralized optimal open-loop control $u^0(\cdot|t_0)$. $\quad\square$

An important consequence of Theorem 3 is that the performance of the overall system under the distributed feedback tends to centralized closed-loop performance under open-loop optimal feedback when $\max_{w\in W} \|w\| \to 0$ for all $i \in I$ and, as a result, $J^0(\tau) \to J^0(t_0)$. Therefore, for small disturbances the distributed scheme produces a suboptimal feedback.

5 Example

As an illustrative example consider an optimal control of five identical point masses moving in the plane under disturbances and coupled only by the constraints. The dynamics of the systems is given by the equations

$$m\ddot{x}_{i,k} + k_1\dot{x}_{i,k} = u_{i,k} + w_{i,k}, \ k = 1,2,$$

where $i \in I = \{1,\ldots,5\}$ and all systems have same parameters: $m = 1$, $k_1 = 1$, $k_2 = 2$. For all $i \in I$ and $t \geq 0$ the input $u_i = (u_{i,1}, u_{i,2})$ has to satisfy a local constraint of the form $\|u_i(t)\|_1 \leq 1$ and the disturbance $w_i = (w_{i,1}, w_{i,2})$ is bounded: $\|w_i(t)\|_\infty \leq w^*$, where $w^* = 0.05$.

The interconnection topology is given by a circle, i.e. every system $i = 2,3,4$ has systems $i + 1$ and $i - 1$ as their neighbors ($N_i = \{i - 1, i + 1\}$) and the first and the last systems are coupled to their closest neighbor and each other: $N_1 = \{2,5\}$, $N_5 = \{4,1\}$.

The control objective is to drive in finite time $t_f = 10$ all systems closer to their neighbors

$$|x_{i,k}(t_f) - x_{j,k}(t_f)| \leq 1, k = 1,2, \ j \in N_i, i \in I,$$

while maximizing the worst-case terminal velocity along the vertical axes:

$$\max_{u_i} \min_{w_i} \dot{x}_{i,2}(t_f), \ i \in I.$$

Figure 1 shows simulation results when applying the distributed Algorithm 2 in comparison with centralized optimal control by Algorithm 1. In the simulation presented at time $t = 0$ all masses were stationary at different positions and for $t \in [0, t_f]$ subjected to the following constant disturbances: $w_1(t) = w^*(-0.5, -1)^T$, $w_2(t) = -w_1(t)$, $w_3(t) = w^*(0.5, 0.5)^T$, $w_4(t) = w^*(0, 1)^T$, $w_5(t) = w^*(0.5, -1)^T$. The weights in (8) were all equal to $1/2$.

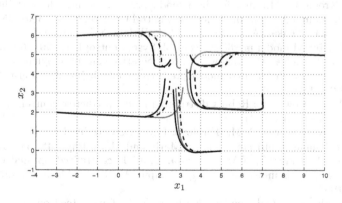

Fig. 1. State trajectories under centralized (solid), distributed (dash) and open-loop (grey) inputs

It can be seen from Fig. 1 that distributed control recovers the behavior of the centralized controller. The performance index of the centralized open-loop optimal feedback was 2.39263225 while the one of the distributed feedback was 2.36410018, which constitutes a loss in performance of only 1.1925 per cent. In other simulations for dynamically coupled [8] and decoupled systems the percentage error was not over five per cent.

For the reference the disturbed optimal open-loop trajectories calculated at the initial time $t = 0$ and used for initialization of the distributed algorithm are also presented in Fig. 1. It can be seen that the distributed trajectories can deviate quite far from the initial centralized plan.

6   Conclusions

This paper presents a robust distributed control scheme for optimal control of linear dynamically decoupled systems with coupling constraints. The key advantages of the algorithm are (1) parallel solutions of local optimization problems without interactions during iterations, (2) small amount of communication data, (3) robust constraint satisfaction, (4) less conservatism due to an assumption of the nominal and not worst-case performance of the neighbor systems. Future research will focus on obtaining suboptimality estimates for robust distributed feedbacks and development of the scheme for weakly interconnected systems.

Acknowledgments. This work is partially supported by the Belarusian Republican Foundation for Fundamental Research (grant Φ14MC-005).

References

1. Scattolini, R.: Architectures for distributed and hierarchical model predictive control - a review. J. of Process Control **19**, 723–731 (2009)
2. Camponogara, E., Jia, D., Krogh, B.H., Talukdar, S.: Distributed model predictive control. IEEE Control Syst. Mag. **22**, 44–52 (2002)
3. Jia, D., Krogh, B.: Min-max feedback model predictive control for distributed control with communication. In: American Control Conference, pp. 4507–4512 (2002)
4. Stewart, B.T., Venkat, A.N., Rawlings, J.B., Wright, S.J., Pannocchia, G.: Cooperative distributed model predictive control. Syst. Control Lett. **59**, 460–469 (2010)
5. Müller, M.A., Reble, M., Allgöwer, F.: Cooperative control of dynamically decoupled systems via distributed model predictive control. Int. J. Robust Nonlinear Control **22**, 1376–1397 (2012)
6. Richards, A., How, J.P.: Robust distributed model predictive control. Int. J. Control **80**, 1517–1531 (2007)
7. Gabasov, R., Dmitruk, N.M., Kirillova, F.M.: Decentralized optimal control of a group of dynamical objects. Comput. Math. and Math. Phys. **48**, 561–576 (2008)
8. Gabasov, R., Dmitruk, N.M., Kirillova, F.M.: Decentralized optimal control of dynamical systems under uncertainty. Comput. Math. and Math. Phys. **51**, 1128–1145 (2011)
9. Balashevich, N.V., Gabasov, R., Kirillova, F.M.: Guaranteed on-line control for linear systems under disturbances. Funct. Diff. Eq. **11**, 341–361 (2004)
10. Kostyukova, O., Kostina, E.: Robust optimal feedback for terminal linear-quadratic conrol problems under disturbances. Math. Program. **107**, 131–153 (2006)

An Optimal Control Approach to Herglotz Variational Problems

Simão P.S. Santos, Natália Martins, and Delfim F.M. Torres[✉]

CIDMA–Center for Research and Development in Mathematics and Applications,
Department of Mathematics, University of Aveiro, 3810-193 Aveiro, Portugal
{spsantos,natalia,delfim}@ua.pt

Abstract. We address the generalized variational problem of Herglotz from an optimal control point of view. Using the theory of optimal control, we derive a generalized Euler–Lagrange equation, a transversality condition, a DuBois–Reymond necessary optimality condition and Noether's theorem for Herglotz's fundamental problem, valid for piecewise smooth functions.

Keywords: Herglotz's variational problems · Optimal control · Euler–Lagrange equations · Invariance · Dubois–Reymond condition · Noether's theorem

1 Introduction

The generalized variational problem proposed by Herglotz in 1930 [3,4] can be formulated as follows:

$$z(b) \longrightarrow \text{extr}$$
$$\text{with } \dot{z}(t) = L(t, x(t), \dot{x}(t), z(t)), \quad t \in [a, b], \qquad (P_H)$$
$$\text{subject to } x(a) = \alpha, \quad z(a) = \gamma, \quad \alpha, \gamma \in \mathbb{R}.$$

It consists in the determination of trajectories $x(\cdot)$ and corresponding trajectories $z(\cdot)$ that extremize (maximize or minimize) the value $z(b)$, where $L \in C^1([a, b] \times \mathbb{R}^{2n} \times \mathbb{R}; \mathbb{R})$. While in [3,4,6] the admissible functions are $x(\cdot) \in C^2([a, b]; \mathbb{R}^n)$ and $z(\cdot) \in C^1([a, b]; \mathbb{R})$, here we consider (P_H) in the wider class of functions $x(\cdot) \in PC^1([a, b]; \mathbb{R}^n)$ and $z(\cdot) \in PC^1([a, b]; \mathbb{R})$.

It is obvious that Herglotz's problem (P_H) reduces to the classical fundamental problem of the calculus of variations (see, e.g., [13]) if the Lagrangian L does not depend on the z variable: if $\dot{z}(t) = L(t, x(t), \dot{x}(t))$, $t \in [a, b]$, then (P_H) is equivalent to the classical variational problem

$$\int_a^b L(t, x(t), \dot{x}(t))dt \longrightarrow \text{extr}, \quad x(a) = \alpha. \qquad (1)$$

Part of first author's Ph.D. project, which is carried out under the *Doctoral Programme in Mathematics* (PDMat) of University of Aveiro.

© Springer International Publishing Switzerland 2015
A. Plakhov et al. (Eds.): EmC-ONS 2014, CCIS 499, pp. 107–117, 2015.
DOI: 10.1007/978-3-319-20352-2_7

Herglotz proved that an Euler–Lagrange optimality condition for a pair $(x(\cdot), z(\cdot))$ to be an extremizer of the generalized variational problem (P_H) is given by

$$
\frac{\partial L}{\partial x}(t, x(t), \dot{x}(t), z(t)) - \frac{d}{dt}\frac{\partial L}{\partial \dot{x}}(t, x(t), \dot{x}(t), z(t))
$$
$$
+ \frac{\partial L}{\partial z}(t, x(t), \dot{x}(t), z(t))\frac{\partial L}{\partial \dot{x}}(t, x(t), \dot{x}(t), z(t)) = 0, \quad (2)
$$

$t \in [a, b]$. The Eq. (2) is known as the generalized Euler–Lagrange equation. Observe that for the fundamental problem of the calculus of variations (1) one has $\frac{\partial L}{\partial z} = 0$ and the differential Eq. (2) reduces to the classical Euler–Lagrange equation

$$
\frac{\partial L}{\partial x}(t, x(t), \dot{x}(t)) - \frac{d}{dt}\frac{\partial L}{\partial \dot{x}}(t, x(t), \dot{x}(t)) = 0.
$$

Since the celebrated work [5] by Pontryagin et al., the calculus of variations is seen as part of optimal control. One of the simplest problems of optimal control, in Bolza form, is the following one:

$$
\mathcal{J}(x(\cdot), u(\cdot)) = \int_a^b f(t, x(t), u(t))dt + \phi(x(b)) \longrightarrow \text{extr} \tag{P}
$$
$$
\text{subject to } \dot{x}(t) = g(t, x(t), u(t)) \text{ and } x(a) = \alpha, \quad \alpha \in \mathbb{R},
$$

where $f \in C^1([a, b] \times \mathbb{R}^n \times \Omega; \mathbb{R})$, $\phi \in C^1(\mathbb{R}^n; \mathbb{R})$, $g \in C^1([a, b] \times \mathbb{R}^n \times \Omega; \mathbb{R}^n)$, $x \in PC^1([a, b]; \mathbb{R}^n)$ and $u \in PC([a, b]; \Omega)$, with $\Omega \subseteq \mathbb{R}^r$ an open set. In the literature of optimal control, x and u are called the state and control variables, respectively, while ϕ is known as the payoff or salvage term. Note that the classical problem of the calculus of variations (1) is a particular case of problem (P) with $\phi(x) \equiv 0$, $g(t, x, u) = u$ and $\Omega = \mathbb{R}^n$. In this work we show how the results on Herglotz's problem of the calculus of variations (P_H) obtained in [2,6] can be generalized by using the theory of optimal control. The main idea is simple and consists in rewriting the generalized variational problem of Herglotz (P_H) as a standard optimal control problem (P), and then to apply available results of optimal control theory.

The paper is organized as follows. In Sect. 2 we briefly review the necessary concepts and results from optimal control theory. In particular, we make use of Pontryagin's maximum principle (Theorem 1); the DuBois–Reymond condition of optimal control (Theorem 2); and the Noether theorem of optimal control proved in [8] (cf. Theorem 3). Our contributions are then given in Sect. 3: we generalize the Euler–Lagrange equation and the transversality condition for problem (P_H) found in [6] to admissible functions $x(\cdot) \in PC^1([a, b]; \mathbb{R}^n)$ and $z(\cdot) \in PC^1([a, b]; \mathbb{R})$ (Theorem 4); we obtain a DuBois–Reymond necessary optimality condition for problem (P_H) (Theorem 5); and a generalization of the Noether theorem [2] (Theorem 6) as a corollary of the optimal control results of Torres [7–9]. We end with Sect. 4 of conclusions and future work.

2   Preliminaries

The central result in optimal control theory is given by Pontryagin's maximum principle, which is a first-order necessary optimality condition.

Theorem 1 (Pontryagin's Maximum Principle for Problem *(P)* [5]**).** *If a pair $(x(\cdot), u(\cdot))$ with $x \in PC^1([a, b]; \mathbb{R}^n)$ and $u \in PC([a, b]; \Omega)$ is a solution to problem (P), then there exists $\psi \in PC^1([a, b]; \mathbb{R}^n)$ such that the following conditions hold:*

– the optimality condition

$$\frac{\partial H}{\partial u}(t, x(t), u(t), \psi(t)) = 0; \tag{3}$$

– the adjoint system

$$\begin{cases} \dot{x}(t) = \frac{\partial H}{\partial \psi}(t, x(t), u(t), \psi(t)) \\ \dot{\psi}(t) = -\frac{\partial H}{\partial x}(t, x(t), u(t), \psi(t)); \end{cases} \tag{4}$$

– and the transversality condition

$$\psi(b) = \nabla\phi(x(b)); \tag{5}$$

where the Hamiltonian H is defined by

$$H(t, x, u, \psi) = f(t, x, u) + \psi \cdot g(t, x, u). \tag{6}$$

Definition 1 (Pontryagin Extremal to *(P)***).** *A triplet $(x(\cdot), u(\cdot), \psi(\cdot))$ with $x \in PC^1([a, b]; \mathbb{R}^n)$, $u \in PC([a, b]; \Omega)$ and $\psi \in PC^1([a, b]; \mathbb{R}^n)$ is called a Pontryagin extremal to problem (P) if it satisfies the optimality condition (3), the adjoint system (4) and the transversality condition (5).*

Theorem 2 (DuBois–Reymond Condition of Optimal Control [5]**).** *If $(x(\cdot), u(\cdot), \psi(\cdot))$ is a Pontryagin extremal to problem (P), then the Hamiltonian (6) satisfies the equality*

$$\frac{dH}{dt}(t, x(t), u(t), \psi(t)) = \frac{\partial H}{\partial t}(t, x(t), u(t), \psi(t)),$$

$t \in [a, b]$.

Noether's theorem has become a fundamental tool of modern theoretical physics [1], the calculus of variations [10,11], and optimal control [7–9]. It states that when an optimal control problem is invariant under a one parameter family of transformations, then there exists a corresponding conservation law: an expression that is conserved along all the Pontryagin extremals of the problem [7–9,12]. Here we use Noether's theorem as found in [8], which is formulated for problems of optimal control in Lagrange form, that is, for problem (P) with $\phi \equiv 0$.

In order to apply the results of [8] to the Bolza problem (P), we rewrite it in the following equivalent Lagrange form:

$$\mathcal{I}(x_0(\cdot), x(\cdot), u(\cdot)) = \int_a^b [f(t, x(t), u(t)) + x_0(t)]\, dt \longrightarrow \text{extr},$$

$$\begin{cases} \dot{x}_0(t) = 0, \\ \dot{x}(t) = g\,(t, x(t), u(t)), \end{cases} \tag{7}$$

$$x_0(a) = \frac{\phi(x(b))}{b-a}, \quad x(a) = \alpha.$$

The notion of invariance for problem (P) is obtained by applying the notion of invariance found in [8] to the equivalent optimal control problem (7). In Definition 2 we use the little-o notation.

Definition 2 (Invariance of Problem (P)). *Let h^s be a one-parameter family of C^1 invertible maps*

$$h^s : [a, b] \times \mathbb{R}^n \times \Omega \to \mathbb{R} \times \mathbb{R}^n \times \mathbb{R}^r,$$
$$h^s(t, x, u) = (T^s(t, x, u), \mathcal{X}^s(t, x, u), \mathcal{U}^s(t, x, u)),$$
$$h^0(t, x, u) = (t, x, u) \text{ for all } (t, x, u) \in [a, b] \times \mathbb{R}^n \times \Omega.$$

Problem (P) is said to be invariant under transformations h^s if for all $(x(\cdot), u(\cdot))$ the following two conditions hold:

(i)

$$\left[f \circ h^s(t, x(t), u(t)) + \frac{\phi(x(b))}{b-a} + \xi s + o(s) \right] \frac{dT^s}{dt}(t, x(t), u(t))$$

$$= f(t, x(t), u(t)) + \frac{\phi(x(b))}{b-a} \quad (8)$$

for some constant ξ;

(ii)

$$\frac{d\mathcal{X}^s}{dt}(t, x(t), u(t)) = g \circ h^s(t, x(t), u(t)) \frac{dT^s}{dt}(t, x(t), u(t)). \tag{9}$$

Theorem 3 (Noether's Theorem for the Optimal Control Problem (P)). *If problem (P) is invariant in the sense of Definition 2, then the quantity*

$$(b-t)\xi + \psi(t) \cdot X(t, x(t), u(t)) - \left[H(t, x(t), u(t), \psi(t)) + \frac{\phi(x(b))}{b-a} \right] \cdot T(t, x(t), u(t))$$

is constant in t along every Pontryagin extremal $(x(\cdot), u(\cdot), \psi(\cdot))$ of problem (P), where

$$T(t, x(t), u(t)) = \left. \frac{\partial T^s}{\partial s}(t, x(t), u(t)) \right|_{s=0},$$

$$X(t, x(t), u(t)) = \left. \frac{\partial \mathcal{X}^s}{\partial s}(t, x(t), u(t)) \right|_{s=0},$$

and H is defined by (6).

Proof. The result is a simple exercise obtained by applying the Noether theorem of [8] and the Pontryagin maximum principle (Theorem 1) to the equivalent optimal control problem (7) (in particular using the adjoint equation corresponding to the multiplier associated with the state variable x_0 and the respective transversality condition).

3   Main Results

We begin by introducing some basic definitions for the generalized variational problem of Herglotz (P_H).

Definition 3 (Admissible Pair to Problem (P_H)). *We say that $(x(\cdot), z(\cdot))$ with $x(\cdot) \in PC^1([a, b]; \mathbb{R}^n)$ and $z(\cdot) \in PC^1([a, b]; \mathbb{R})$ is an admissible pair to problem (P_H) if it satisfies the equation*

$$\dot{z}(t) = L(t, x(t), \dot{x}(t), z(t)), \quad t \in [a, b],$$

and the initial conditions $x(a) = \alpha$ and $z(a) = \gamma$, $\alpha, \gamma \in \mathbb{R}$.

Definition 4 (Extremizer to Problem (P_H)). *We say that an admissible pair $(x^*(\cdot), z^*(\cdot))$ is an extremizer to problem (P_H) if $z(b) - z^*(b)$ has the same signal for all admissible pairs $(x(\cdot), z(\cdot))$ that satisfy $\|z - z^*\|_0 < \epsilon$ and $\|x - x^*\|_0 < \epsilon$ for some positive real ϵ, where $\|y\|_0 = \max_{a \le t \le b} |y(t)|$.*

We now present a necessary condition for a pair $(x(\cdot), z(\cdot))$ to be a solution (extremizer) to problem (P_H). The following result generalizes [3,4,6] by considering a more general class of functions. To simplify notation, we use the operator $\langle \cdot, \cdot \rangle$ defined by

$$\langle x, z \rangle(t) := (t, x(t), \dot{x}(t), z(t)).$$

When there is no possibility of ambiguity, we sometimes suppress arguments.

Theorem 4 (Euler–Lagrange Equation and Transversality Condition for Problem (P_H)). *If $(x(\cdot), z(\cdot))$ is an extremizer to problem (P_H), then the Euler–Lagrange equation*

$$\frac{\partial L}{\partial x} \langle x, z \rangle(t) - \frac{d}{dt} \left(\frac{\partial L}{\partial \dot{x}} \right) \langle x, z \rangle(t) + \frac{\partial L}{\partial z} \langle x, z \rangle(t) \frac{\partial L}{\partial \dot{x}} \langle x, z \rangle(t) = 0 \qquad (10)$$

holds, $t \in [a, b]$. Moreover, the following transversality condition holds:

$$\frac{\partial L}{\partial \dot{x}} \langle x, z \rangle(b) = 0. \qquad (11)$$

Proof. Observe that Herglotz's problem (P_H) is a particular case of problem (P) obtained by considering x and z as state variables (two components of one vectorial state variable), $\dot{x}$ as the control variable u, and by choosing $f \equiv 0$ and $\phi(x, z) = z$. Note that since $x(t) \in \mathbb{R}^n$, we have $u(t) \in \mathbb{R}^n$ (i.e., for Herglotz's problem (P_H) one has $r = n$). In this way, the problem of Herglotz, described as an optimal control problem, takes the form

$$z(b) \longrightarrow \text{extr},$$

$$\begin{cases} \dot{x}(t) = u(t), \\ \dot{z}(t) = L(t, x(t), u(t), z(t)), \end{cases} \tag{12}$$

$$x(a) = \alpha, \quad z(a) = \gamma, \quad \alpha, \gamma \in \mathbb{R}.$$

It follows from Pontryagin's maximum principle (Theorem 1) that there exists $\psi_x \in PC^1([a, b]; \mathbb{R}^n)$ and $\psi_z \in PC^1([a, b]; \mathbb{R})$ such that the following conditions hold:

– the optimality condition

$$\frac{\partial H}{\partial u}(t, x(t), u(t), z(t), \psi_x(t), \psi_z(t)) = 0; \tag{13}$$

– the adjoint system

$$\begin{cases} \dot{x}(t) = \frac{\partial H}{\partial \psi_x}(t, x(t), u(t), z(t), \psi_x(t), \psi_z(t)) \\ \dot{z}(t) = \frac{\partial H}{\partial \psi_z}(t, x(t), u(t), z(t), \psi_x(t), \psi_z(t)) \\ \dot{\psi}_x(t) = -\frac{\partial H}{\partial x}(t, x(t), u(t), z(t), \psi_x(t), \psi_z(t)) \\ \dot{\psi}_z(t) = -\frac{\partial H}{\partial z}(t, x(t), u(t), z(t), \psi_x(t), \psi_z(t)); \end{cases} \tag{14}$$

– and the transversality conditions

$$\begin{cases} \psi_x(b) = 0, \\ \psi_z(b) = 1, \end{cases} \tag{15}$$

where the Hamiltonian H is defined by

$$H(t, x, u, z, \psi_x, \psi_z) = \psi_x \cdot u + \psi_z \cdot L(t, x, u, z).$$

Observe that the adjoint system (14) implies that

$$\begin{cases} \dot{\psi}_x = -\psi_z \frac{\partial L}{\partial x} \\ \dot{\psi}_z = -\psi_z \frac{\partial L}{\partial z}. \end{cases} \tag{16}$$

This means that ψ_z is solution of a first-order linear differential equation, which is solved using an integrand factor to find that $\psi_z = ke^{-\int_a^t \frac{\partial L}{\partial z} d\theta}$ with k a constant. From the second transversality condition in (15), we obtain that $k = e^{\int_a^b \frac{\partial L}{\partial z} d\theta}$ and, consequently,

$$\psi_z = e^{\int_t^b \frac{\partial L}{\partial z} d\theta}.$$

The optimality condition (13) is equivalent to $\psi_x + \psi_z \frac{\partial L}{\partial u} = 0$ and, after derivation, we obtain that

$$\dot{\psi}_x = -\frac{d}{dt}\left(\psi_z \frac{\partial L}{\partial u}\right) = -\dot{\psi}_z \frac{\partial L}{\partial u} - \psi_z \frac{d}{dt}\left(\frac{\partial L}{\partial u}\right) = \psi_z \frac{\partial L}{\partial z} \frac{\partial L}{\partial u} - \psi_z \frac{d}{dt}\left(\frac{\partial L}{\partial u}\right).$$

Now, comparing with (16), we have

$$-\psi_z \frac{\partial L}{\partial x} = \psi_z \frac{\partial L}{\partial z} \frac{\partial L}{\partial u} - \psi_z \frac{d}{dt}\left(\frac{\partial L}{\partial u}\right).$$

Since $\psi_z(t) \neq 0$ for all $t \in [a,b]$ and $\dot{x} = u$, we obtain the Euler–Lagrange Eq. (10):

$$\frac{\partial L}{\partial x} - \frac{d}{dt}\left(\frac{\partial L}{\partial \dot{x}}\right) + \frac{\partial L}{\partial z}\frac{\partial L}{\partial \dot{x}} = 0.$$

Note that from the optimality condition (13) we obtain that $\psi_x = -\psi_z \frac{\partial L}{\partial u} = -\psi_z \frac{\partial L}{\partial \dot{x}}$, which together with transversality condition (15) for ψ_x leads to the transversality condition (11):

$$\frac{\partial L}{\partial \dot{x}}(b, x(b), \dot{x}(b), z(b)) = 0.$$

This concludes the proof.

Definition 5 (Extremal to Problem (P_H)). *We say that an admissible pair $(x(\cdot), z(\cdot))$ is an extremal to problem (P_H) if it satisfies the Euler–Lagrange Eq. (10) and the transversality condition (11).*

Theorem 5 (DuBois–Reymond Condition for Problem (P_H)). *If $(x(\cdot), z(\cdot))$ is an extremal to problem (P_H), then*

$$\frac{d}{dt}\left(-\psi_z(t)\frac{\partial L}{\partial \dot{x}}\langle x, z\rangle(t)\dot{x}(t) + \psi_z(t)L\langle x, z\rangle(t)\right) = \psi_z(t)\frac{\partial L}{\partial t}\langle x, z\rangle(t),$$

$t \in [a,b]$, where $\psi_z(t) = e^{\int_t^b \frac{\partial L}{\partial z}\langle x, z\rangle(\theta)d\theta}$.

Proof. The result follows from Theorem 2, rewriting problem (P_H) as the optimal control problem (12). $\qquad\blacksquare$

We define invariance for (P_H) using Definition 2 for the equivalent optimal control problem (12).

Definition 6 (Invariance of Problem (P_H)). *Let h^s be a one-parameter family of C^1 invertible maps*

$$h^s : [a,b] \times \mathbb{R}^n \times \mathbb{R} \to \mathbb{R} \times \mathbb{R}^n \times \mathbb{R},$$
$$h^s(t, x(t), z(t)) = (\mathcal{T}^s\langle x, z\rangle(t), \mathcal{X}^s\langle x, z\rangle(t), \mathcal{Z}^s\langle x, z\rangle(t)),$$
$$h^0(t, x, z) = (t, x, z), \quad \forall(t, x, z) \in [a,b] \times \mathbb{R}^n \times \mathbb{R}.$$

Problem (P_H) is said to be invariant under the transformations h^s if for all admissible pairs $(x(\cdot), z(\cdot))$ the following two conditions hold:

(i)

$$\left(\frac{z(b)}{b-a} + \xi s + o(s)\right)\frac{d\mathcal{T}^s}{dt}\langle x, z\rangle(t) = \frac{z(b)}{b-a} \tag{17}$$

for some constant ξ;

(ii)

$$\frac{d\mathcal{Z}^s}{dt}\langle x, z\rangle(t)$$

$$= L\left(T^s\langle x, z\rangle(t), \mathcal{X}^s\langle x, z\rangle(t), \frac{d\mathcal{X}^s}{dT^s}\langle x, z\rangle(t), \mathcal{Z}^s\langle x, z\rangle(t)\right)\frac{dT^s}{dt}\langle x, z\rangle(t),$$

$$(18)$$

where

$$\frac{d\mathcal{X}^s}{dT^s}\langle x, z\rangle(t) = \frac{\frac{d\mathcal{X}^s}{dt}\langle x, z\rangle(t)}{\frac{dT^s}{dt}\langle x, z\rangle(t)}.$$

Follows the main result of the paper.

Theorem 6 (Noether's Theorem for Problem (P_H)). *If problem (P_H) is invariant in the sense of Definition 6, then the quantity*

$$\psi_z(t)\left[\frac{\partial L}{\partial \dot{x}}\langle x, z\rangle(t)X\langle x, z\rangle(t) - Z\langle x, z\rangle(t)\right.$$

$$\left. + \left(L\langle x, z\rangle(t) - \frac{\partial L}{\partial \dot{x}}\langle x, z\rangle(t)\dot{x}(t)\right)T\langle x, z\rangle(t)\right] \quad (19)$$

is constant in t along every extremal of problem (P_H), where

$$T\langle x, z\rangle(t) = \left.\frac{\partial T^s}{\partial s}\langle x, z\rangle(t)\right|_{s=0},$$

$$X\langle x, z\rangle(t) = \left.\frac{\partial \mathcal{X}^s}{\partial s}\langle x, z\rangle(t)\right|_{s=0},$$

$$Z\langle x, z\rangle(t) = \left.\frac{\partial \mathcal{Z}^s}{\partial s}\langle x, z\rangle(t)\right|_{s=0}$$

and $\psi_z(t) = e^{\int_t^b \frac{\partial L}{\partial z}\langle x, z\rangle(\theta)d\theta}$.

Proof. As before, we rewrite problem (P_H) in the equivalent optimal control form (12), where x and z are the state variables and u the control. We prove that if problem (P_H) is invariant in the sense of Definition 6, then (12) is invariant in the sense of Definition 2. First, observe that if Eq. (17) holds, then (8) holds for (12): here $f \equiv 0$, $\phi(x, z) = z$ and (8) simplifies to $\left[\frac{z(b)}{b-a} + \xi s + o(s)\right]\frac{dT^s}{dt}\langle x, z\rangle(t) = \frac{z(b)}{b-a}$. Note that the first equation of the control system of problem (12) ($u(t) = \dot{x}(t)$) defines $\mathcal{U}^s := \frac{d\mathcal{X}^s}{dT^s}$, that is,

$$\frac{d\mathcal{X}^s}{dt}\langle x, z\rangle(t) = \mathcal{U}^s\langle x, z\rangle(t)\frac{dT^s}{dt}\langle x, z\rangle(t). \quad (20)$$

Hence, if Eqs. (18) and (20) holds, then there is also invariance of the control system of (12) in the sense of (9) and consequently problem (12) is invariant in the sense of Definition 2. We are now in conditions to apply Theorem 3 to problem (12), which guarantees that the quantity

$$(b - t)\xi + \psi_x(t) \cdot X(t, x(t), u(t), z(t)) + \psi_z(t) \cdot Z(t, x(t), u(t), z(t))$$
$$- \left(H(t, x(t), u(t), z(t), \psi_x(t), \psi_z(t)) + \frac{z(b)}{b - a} \right) \cdot T(t, x(t), u(t), z(t))$$

is constant in t along every Pontryagin extremal of problem (12), where

$$H(t, x, u, z, \psi_x, \psi_z) = \psi_x u + \psi_z L(t, x, u, z).$$

This means that the quantity

$$(b - t)\xi + \psi_x(t)X\langle x, z\rangle(t) + \psi_z(t)Z\langle x, z\rangle(t)$$
$$- \left(\psi_x(t)\dot{x}(t) + \psi_z(t)L\langle x, z\rangle(t) + \frac{z(b)}{b - a} \right) T\langle x, z\rangle(t)$$

is constant in t along all extremals of problem (P_H), where

$$\psi_x(t) = -\psi_z(t)\frac{\partial L}{\partial u}\langle x, z\rangle(t) = -\psi_z(t)\frac{\partial L}{\partial \dot{x}}\langle x, z\rangle(t).$$

Equivalently,

$$(b - t)\xi - \frac{z(b)}{b - a}T\langle x, z\rangle(t) - \psi_z(t)\left[\frac{\partial L}{\partial \dot{x}}\langle x, z\rangle(t)X\langle x, z\rangle(t) - Z\langle x, z\rangle(t) \right.$$
$$\left. + \left(L\langle x, z\rangle(t) - \frac{\partial L}{\partial \dot{x}}\langle x, z\rangle(t)\dot{x}(t) \right) T\langle x, z\rangle(t) \right]$$

is a constant along the extremals. To conclude the proof, we just need to prove that the quantity

$$(b - t)\xi - \frac{z(b)}{b - a}T\langle x, z\rangle(t) \tag{21}$$

is a constant. From the invariance condition (17) we know that

$$(z(b) + \xi(b - a)s + o(s))\frac{dT^s}{dt}\langle x, z\rangle(t) = z(b).$$

Integrating from a to t, we conclude that

$$\left(z(b) + \xi(b - a)s + o(s) \right)T^s\langle x, z\rangle(t)$$
$$= z(b)(t - a) + (z(b) + \xi(b - a)s + o(s))T^s\langle x, z\rangle(a). \tag{22}$$

Differentiating (22) with respect to s, and then putting $s = 0$, we obtain

$$\xi(b - a)t + z(b)T\langle x, z\rangle(t) = \xi(b - a)a + z(b)T\langle x, z\rangle(a). \tag{23}$$

We conclude from (23) that expression (21) is the constant $(b - a)\xi - \frac{z(b)}{b-a} T\langle x, z\rangle(a)$.

4 Conclusion

We introduced a different approach to the generalized variational principle of Herglotz, by looking to Herglotz's problem as an optimal control problem. A Noether type theorem for Herglotz's problem was first proved by Georgieva and Guenther in [2]: under the condition of invariance

$$\frac{d}{ds}\left[L\left(T^s\langle x,z\rangle(t), \mathcal{X}^s\langle x,z\rangle(t), \frac{d\mathcal{X}^s}{dT^s}\langle x,z\rangle(t), z(t)\right)\frac{dT^s}{dt}\langle x,z\rangle(t)\right]\Bigg|_{s=0} = 0,$$
(24)

they obtained

$$\lambda(t)\left[\frac{\partial L}{\partial \dot{x}}\langle x,z\rangle(t)X\langle x,z\rangle(t) + \left(L\langle x,z\rangle(t) - \frac{\partial L}{\partial \dot{x}}\langle x,z\rangle(t)\dot{x}(t)\right)T\langle x,z\rangle(t)\right], \quad (25)$$

where $\lambda(t) = e^{-\int_a^t \frac{\partial L}{\partial z}\langle x,z\rangle(\theta)d\theta}$, as a conserved quantity along the extremals of problem (P_H). Our results improve those of [2] in three ways: (i) we consider a wider class of piecewise admissible functions; (ii) we consider a more general notion of invariance whose transformations T^s, $\mathcal{X}^s$ and $\mathcal{Z}^s$ may also depend on velocities, i.e., on $\dot{x}(t)$ (note that if (18) holds with $\mathcal{Z}^s\langle x,z\rangle = z$, then (24) also holds); (iii) the conserved quantity (25), up to multiplication by a constant, is a particular case of (19) when there is no transformation in z ($Z = \frac{\partial \mathcal{Z}^s}{\partial s}\big|_{s=0} = 0$). The results here obtained can be generalized to higher-order variational problems of Herglotz type. This is under investigation and will be addressed elsewhere.

Acknowledgments. This work was supported by Portuguese funds through the *Center for Research and Development in Mathematics and Applications* (CIDMA), within project UID/MAT/04106/2013, and the *Portuguese Foundation for Science and Technology* (FCT). The authors would like to thank an anonymous Reviewer for valuable comments.

References

1. Frederico, G.S.F., Torres, D.F.M.: Fractional isoperimetric Noether's theorem in the Riemann-Liouville sense. Rep. Math. Phys. **71**(3), 291–304 (2013)
2. Georgieva, B., Guenther, R.: First Noether-type theorem for the generalized variational principle of Herglotz. Topol. Methods Nonlinear Anal. **20**(2), 261–273 (2002)
3. Guenther, R.B., Guenther, C.M., Gottsch, J.A.: The Herglotz Lectures on Contact Transformations and Hamiltonian Systems. Lecture Notes in Nonlinear Analysis. Juliusz Schauder Center for Nonlinear Studies, Nicholas Copernicus University, Torún (1996)
4. Herglotz, G.: Berührungstransformationen. Lectures at the University of Göttingen, Göttingen (1930)
5. Pontryagin, L.S., Boltyanskii, V.G., Gamkrelidze, R.V., Mishchenko, E.F.: The Mathematical Theory of Optimal Processes. Interscience Publishers, London (1962)

6. Santos, S.P.S., Martins, N., Torres, D.F.M.: Higher-order variational problems of Herglotz type. Vietnam J. Math. **42**(4), 409–419 (2014)
7. Torres, D.F.M.: On the Noether theorem for optimal control. Eur. J. Control **8**(1), 56–63 (2002)
8. Torres, D.F.M.: Conservation laws in optimal control. In: Colonius, F., Grüne, L. (eds.) Dynamics, Bifurcations, and Control. Lecture Notes in Control and Information Science, vol. 273, pp. 287–296. Springer, Berlin (2002)
9. Torres, D.F.M.: Quasi-invariant optimal control problems. Port. Math. **61**(1), 97–114 (2004). (N.S.)
10. Torres, D.F.M.: Carathéodory equivalence Noether theorems, and tonelli full-regularity in the calculus of variations and optimal control. J. Math. Sci. **120**(1), 1032–1050 (2004). (N. Y.)
11. Torres, D.F.M.: Proper extensions of Noether's symmetry theorem for nonsmooth extremals of the calculus of variations. Commun. Pure Appl. Anal. **3**(3), 491–500 (2004)
12. Torres, D.F.M.: A Noether theorem on unimprovable conservation laws for vector-valued optimization problems in control theory. Georgian Math. J. **13**(1), 173–182 (2006)
13. van Brunt, B.: The Calculus of Variations. Universitext, New York (2004)

Lowering Toxic Concentrations
in the Diesel Exhaust Gases

Elena Pervukhina$^{(\boxtimes)}$, Kostiantyn Osipov, and Victoria Golikova

Sevastopol State University, Universitetskaya Str., 33,
299053 Sevastopol, Russia
elena@pervuh.sebastopol.ua, v.golikova@ua.fm

Abstract. The paper continues the research of the applied optimization problem of the lowering toxic concentrations in the exhaust gases of diesel. The solution of the problem is based on multivariate statistical modeling and optimization technique. The novelty of the approach is in formation of multi-objective function connecting the concentrations of the toxic components with diesel mode parameters on the basis of the vector autoregression model. The advance angle of fuel injection is considered to be the main mode parameter to control toxicants. Taking this into account the multi-objective function is reduced to the scalar objective function.

Keywords: Optimization · Vector autoregression · Diesel exhaust gas · Advance angle of fuel injection

1 Introduction

The paper continues the research of the problem of the lowering toxic concentrations in the exhaust gas of diesel which was mentioned in [1]. The solution of the problem is based on modeling and optimization technique. The special algorithm is created to calculate optimal values of the parameters, determining the minimal toxic concentrations in the exhaust gases and consumed fuel, at different diesel regimes. The realization of the calculated parameters provides the characteristics of cyclic fuel feed and advance angle of fuel injection that allow to improve the economical characteristics of diesels and to decrease their toxicity. This way is considered to be important and economical compared with others (for instance, constructive).

For diesels, the lowering toxicity and smoke without increasing fuel consumption can be only achieved on the basis of the optimal systems for exhaust gas recirculation, injection of the water into intake manifold, optimal design of the fuel system elements and fuel system adjustments [2]. Last component of the solution is the most perspective and important, because it can be implemented not only for new diesels, but also for exploited ones. In addition, this solution does not require additional cost to upgrade technical production and capacity of enterprises.

In the earlier publications [2, 3] it was established that it is possible to optimize diesel working processes by optimizing fuel system adjustments, including the injection control system, and models reflecting the relationships between the advances angle of fuel injection, injection pressure, injection timing, as well as parameters

© Springer International Publishing Switzerland 2015
A. Plakhov et al. (Eds.): EmC-ONS 2014, CCIS 499, pp. 118–128, 2015.
DOI: 10.1007/978-3-319-20352-2_8

characterizing the toxicity of diesel and its mode. But the question of choice of the most suitable objective function and its structure still exists.

The paper continues the study of the problem of the lowering toxic concentrations in the exhaust gas of diesel through calculating the values of the fuel injection advance angle, minimizing the concentrations of toxic components in the exhaust gases without increasing fuel consumption. In [1] the values of the parameters were calculated through two optimizing procedures. The first one was to define more exact concentration values of the main toxic components in the exhaust gases by using their dependence on the mode parameters. The quality criterion was the Kulback-Leibler information divergence between distribution parameters of estimations and real variables. The second optimizing procedure was directly used to calculate the advance angle of fuel injection which corresponds to the minimum values of toxic concentrations. Here we are constructing more complicated multiplicative and multi-objective function. It uses the principal of relative compensation and is connected with the concentrations of the major toxic components in diesel exhaust gases. Then we develop the strategy of control of the fuel combustion in the combustion chamber, providing the minimum value of the objective function. The novelty of the approach is also in the empirical relationships between non-stationary random processes, which reflect the changes in the concentration of toxic components in the combustion products, and mode parameters.

The paper is organized as follows. In the second section we study the applied physical problem namely an opportunity to reduce the concentration by changing the mode parameters and adjusting the advanced angle of injection that is by control of the combusting. The third section is devoted to constructing the objective function. In the fourth section the numerical example is considered.

2 The Analysis of the Methods to Lower Toxicity

The composition of the exhaust gases of diesel depends not only on the type of fuel used, but also on the type of organization and efficiency of the diesel workflow. Volume concentration of toxic substances in the exhaust gas is relatively small (0.2... 2 %). Five main components: NOx, CO, CHx, aldehydes $RCHO$, and sulfur dioxide SO_2 – comprise 80... 95 % of the total mass of toxic exhaust gas components. The 3 of these components are subject to current regulations.

Comparison of the relative aggressiveness of the exhaust gas constituents (the aggressiveness of carbon monoxide is a unit of aggressiveness) [4, 5]:

$CO : CnHm : SO_x: NO_x: C: RCHO: C_{20}H_{12} = 1 : 3,16 : 16,5 : 41,1 : 41,1 : 41,5 : 1260000,0$.

By aggressiveness of separate toxic component we mean the degree of its negative influence on biomolecules of the cell membranes of a human being, causing various diseases [5].

The main reason for the formation of CO in the diesel combustion chamber is unequal distribution of fuel in the combustion zone which leads to the emergence of separate areas with low ratio of air excess, where part of the fuel has not been burnt. In these local areas volume concentration of CO can reach 5... 6 %. Another source of

CO is high-temperature zones of a fuel plume, in which the chemical equilibrium is shifted towards the dissociation of carbon dioxide CO_2 with forming *CO* and O_2.

One of the main causes of *CH* formation is the availability of the cold wall surface layers in the combustion chamber. During the combustion the flame spreads to the wall, from which the heat is taken away, and formed radicals are arranged on the cold walls. Thus, light hydrocarbons are constituted from the radicals of not completely combusted fuel in the wall cold layers of cylinder which is of 0.005... 0.3 mm thick [2]. Another reason of *CH* formation is the availability of zones with low ratio of air excess, mainly of the core zone and zone of the torch flame in combustion chamber.

Solid particles mainly consist of soot, metal oxides, sulfates, and water, as well as not burnt fuel parts and the engine oil [4]. Soot, in turn, is mainly composed of carbon *C* (95... 98 %) and chemically bound hydrogen H (1... 3 %).

One of the main methods to reduce the concentration of the toxic components in the exhaust gas is adjusting the advance angle of the start timing for the fuel injection (θ). This is because the change in the advance angle of fuel injection causes the change of the period for the ignition delay. The changing of period of ignition delay changes the rate of the cylinder pressure, heat release rate in the beginning of combustion and hence the maximum temperature gas in the combustion chamber [6]. Decreasing the advanced angle of fuel injection results in lowering the maximum temperature in the combustion chamber and reduces the nitrogen oxides emissions. However, the reduction of nitrogen oxides takes place only in a certain range since the reduction of the advance angle of fuel injection simultaneously increases the opacity of diesel and reduces its fuel economy.

The influence of the advance angle of fuel injection on the toxicity and fuel efficiency mostly appears when diesel functions at the modes with low ratio of air excess. This is due to the fact that for the modes with a higher ratio of air excess, the time taken for the combustion process is sufficient for complete combustion. Therefore, at part-load mode there is an opportunity to reduce the advance angle of fuel injection in order to lower the concentration of nitrogen oxides in exhaust gas without the increase of fuel consumption and opacity [8]. In other words, for each mode of diesel there is the optimal value of the advance angle of fuel injection which corresponds to the minimum of toxicity and maximum of fuel efficiency.

It is known [2, 5, 7] that, for example, for the diesel D-240 the decrease of the advance angle of fuel injection by 6-8 degrees of rotation of crankshaft with respect to its nominal value allows to reduce the NO_x concentration by two times. The increase of the solid particle concentration by 1.6 − 0.3 times and reducing the fuel efficiency by 5 − 3 % are also observed.

3 Optimization Criterion

Considering that the aim of the research is to lower exhaust gases' toxicity without fuel consumption increase, we formalize the problem to be solved in terms of multiobjective optimization:

$$f_1(x) \rightarrow \min_{x \in D}, \quad f_2(x) \rightarrow \min_{x \in D}, \quad \ldots f_m(x) \rightarrow \min_{x \in D}, \tag{1}$$

where m is a number of toxic parameters, $x_t = (Speed_t, Load_t, \theta_t)^T$ is a vector of parameters characterizing the operating mode, *Speed* denotes a rotational speed of crankshaft, *Load* indicates a load applied to crankshaft, D indicates a set of possible values of x_t.

The restriction of the consumed fuel volume is entered. The maximal increasing the specific fuel consumption should not exceed 3–5 % of the initial or maximal permissible fuel consumption which is specified in the technical conditions.

The best solution of the problem of lowering the diesel toxicity through the selection of the optimal values of the advance angle of fuel injection for the given mode parameters is a vector x^*, for which the conditions (1) are performed for all functions $f_j(x)$, $j = 1 \ldots m$ simultaneously.

Each function has a minimum corresponding to different values of the vector x. However, in practice such solution cannot be obtained. Each function reaches extreme for single value x. It is impossible to find x^*, for which the conditions (1) are carried out for all objective functions simultaneously. The way is to find such solution x^{**}, for which the rational compromise of given goals (1) is provided. Multi-objective optimization problem is reduced to a typical problem with one criterion.

Based on the principle of relative compensation we use generalized multiplicative criterion

$$K(x) = \prod_{i=1}^{m} f_i(x). \tag{2}$$

Disparities of the partial criteria in the expression (2) can be introduced through weights λ_i. Then the criterion (2) takes the form

$$K(x) = \prod_{i=1}^{m} f_i^{\lambda_i}(x). \tag{3}$$

It is assumed that the coefficients λ_i, reflecting the importance of appropriate objective function (partial criterion), are normalized and meet the condition

$$\sum_{j=1}^{3} \lambda_j = 1.$$

Numerical values of the coefficients λ_i are stated by experts on the basis of the significance level of the objective functions for a partial model regarding the existing ecological standards.

The novelty is in determination of functions $f_i(x)$ connecting toxic concentrations with mode parameters. We explain the procedure in the next section.

4 The Vector Autoregression Model Based on Empirical Information

We utilize empirical information describing the test results of the low-power single-cylinder four-stroke injection diesel used to drive the generators from [9] and firstly analyzed by authors in [6]. The rated power of diesel is 3.5 kW. Three time series with 48 values characterize the changes of the three parameters of the diesel modes at the rotational speed of the crankshaft 3000 rpm (revolutions per minute). These are the crankshaft rotational speed, denoted like Speed, the antitorque moment, Load, the advance angle of fuel injection θ. Three time series with the same number of values characterize the changes of diesel toxicity: concentrations of hydrocarbons CH, nitrogen oxides NOx, and particulates with the main component carbon-black Smoke. The time series are plotted on Fig. 1.

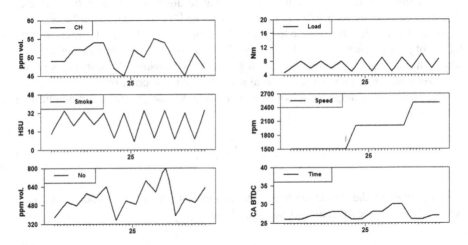

Fig. 1. Time series under studied

At any time the first three parameters form a vector $u_t = (n_e \quad Mc \quad \tau)$ of mode parameters, the latter three parameters form the state vector of the system being studied $x_t = (CH_t, \quad NOx_t, \quad Smoke_t)^T$.

The state vector values for each realization of the vector of mode parameters for the same fuel specification and the same adjustments of the fuel equipment (fuel pressure and injection timing) at each time moment are defined by the advance angle of fuel injection. Based on this fact we formalize the following aim.

It is necessary to determine the values of advance angle of fuel injection which provide the minimum of criterion (3) for given values of the vector of mode parameters. For the example considered the criterion (3) is as follows:

$$K(u) = \frac{NOx(u)^{0.5}}{[NOx]} \cdot \frac{CH(u)^{0.3}}{[CH]} \cdot \frac{Smoke(u)^{0.2}}{[Smoke]} \rightarrow \min. \tag{4}$$

In the Eq. (4) any degree having a meaning of the weight coefficient λ_i has been determined by the experts on the basis of the aggressive importance of the partial component. Symbol [*] denotes the maximum acceptable value. The minimization of the diesel toxicity through the adjustment of the advance angle of fuel injection θ depends on the increase of the average effective specific fuel consumption ge, the values of θ should be selected only from the acceptable range $[\theta_{min} \cdots \theta_{max}]$.

In previous papers, for example [8], it is stated that the dependences $NOx(u), CH(u), Smoke(u)$ with sufficient accuracy (modeling errors are less than 15 – 10 %) can be described by the regression equations:

$$NOx_t = c_1 + \beta_o \cdot NOx_{t-1} + \beta_1 \cdot Speed_t + \beta_2 \cdot Load_t + \beta_3 \cdot \theta_t + \varepsilon_t \qquad (5)$$

$$CH_t = c_2 + \phi_0 \cdot CH_{t-1} + \phi_1 \cdot Speed_t + \phi_2 \cdot Load_t + \phi_3 \cdot \theta_t + \varepsilon_t \qquad (6)$$

$$Smoke_t = c_3 + \xi_0 \cdot Smoke_{t-1} + \xi_1 \cdot Speed_t + \xi_2 \cdot Load_t + \xi_3 \cdot \theta_t + \varepsilon_t \qquad (7)$$

Here β, c, ϕ, and ς are the model coefficients.
For the considered diesel the Eqs. (5)–(7) are the following

$$NOx_t = -0,152NOx_{t-1} - 0,0119 \cdot Speed_t + 40,82 \cdot Load_t + 71,50 \cdot \theta_t + 1601,48 + \varepsilon_t \qquad (8)$$

$$CH_t = -0,0755 \cdot CH_{t-1} + 12.20 - 0,0276 \cdot Speed_t + 0,523 \cdot Load_t + 1.882 \cdot \theta_t + \varepsilon_t \qquad (9)$$

$$Smoke_t = -0,023 \cdot Smoke_{t-1} + 2.5 - 0,015 \cdot Speed_t + 6,49 \cdot Load_t + 0,078 \cdot \theta_t + \varepsilon_t \qquad (10)$$

The models (8)–(10) describe the changes of the diesel toxicity indicators in the vicinity of the working points with some limitations. Models (8)–(10) can be used only to describe the parameters in the vicinity of the working points selected in advance. In this case, the deviations from the working points should not exceed 10 % which makes it difficult to use the above mentioned models to describe the changes in all required velocities of the crankshaft rotation and loads.

In other words, when we use the models (8)–(10) we have to calculate the coefficients at least for 10 points every 30-35 s^{-1} that leads to the necessity to utilize at least 30 equations to describe dependencies for the most probable modes of the diesel. The models (8)–(10) do not take into account the non-stationary character of the random processes reflecting the changes of the mode parameters.

The random nature of the parameters is explained by functional dependency of their values from the values of structural parameters. The structural parameters, in turn, are random that are in the so-called tolerance extent due to location errors and deviations from specification parameters of the cutting part of the cutting tool, the presence of undesired vibration of metal working equipment, etc. [9].

To overcome these drawbacks, numerical values of the coefficients of Eqs. (8)–(10) were identified by multivariate analysis [9, 10].

Different parameters characterizing diesel toxicity are united into the vector. The vector autoregression model is built up [10]:

$$x_t = A_0 + \sum_{j=1}^{p} A_j x_{t-j} + \varepsilon_j, \quad t = 1, \ldots, T \tag{11}$$

where $x_t = (X_{1,t}, \ldots, X_{n,t})^T$ is the vector of parameters, A_0 is $(n \times 1)$-matrix of the coefficients, A_0 is constant $(n \times 1)$-matrix. $\varepsilon_t = (\varepsilon_{1t}, \ldots, \varepsilon_{nt})^T$ is the vector of estimation errors. n is a number of parameters, p is the model order. For example, for the function $NOx = f_1(Speed, Load, \theta)$ reflected by the model (8) the autoregression model (11) can be written as:

$$\begin{bmatrix} NOx \\ Speed \\ Load \\ \theta \end{bmatrix} = \begin{bmatrix} 806 \\ -1171 \\ 4.45 \\ 8.15 \end{bmatrix} + \begin{bmatrix} 1.05 & -0.045 & -2.48 & -38.5 \\ -0.065 & 1.07 & -32.95 & 68.6 \\ 0 & 0 & 0.51 & -0.077 \\ 0.001 & 0 & 0.14 & 0.63 \end{bmatrix} \begin{bmatrix} NOx_{t-1} \\ Speed_{t-1} \\ Load_{t-1} \\ \theta_{t-1} \end{bmatrix} + \varepsilon_t \tag{12}$$

Coefficients are calculated by the least square method. To estimate modeling residuals (Fig. 2) we use the software RATS [11].

To determine the cause and effect relationships (possible linear combinations) between the studied parameters we use the vector autoregression model for differences of the initial non-stationary processes of the parameter changes:

$$\Delta x_t = \alpha + \mu t + \Pi x_{t-1} + \sum_{j=1}^{p-1} \Gamma_j \Delta x_{t-j} + \varepsilon_j, \quad t = 1, \ldots, T \tag{13}$$

where Δ the difference operator: $\Delta x_t = x_t - x_{t-1}$,

$$\Pi = I - A_1 - A_2 - \ldots - A_p,$$

$$\Gamma_j = - \sum_{t=j+1}^{p} \Pi_i.$$

Due to the Granger theorem rank r of the matrix $\Pi = \alpha \beta'$ equals the number of linear dependencies [9]. The rows of the matrix β' represent the different vectors, reflecting stable statistical relationships between the studied parameters.

For parameter vector $X_t = [NOx \quad Speed \quad Load \quad \theta]^T$ the matrix $\Pi = \alpha \beta^T$ has the following form:

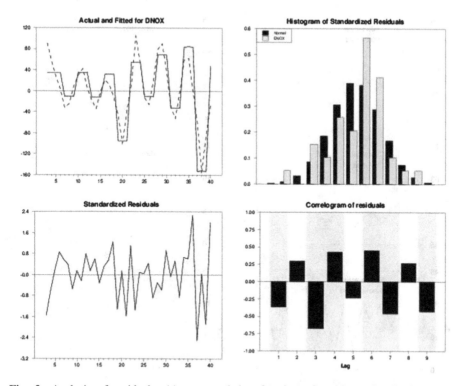

Fig. 2. Analysis of residuals: (a) autocorrelation function of residues; (b) distribution of residuals, 1 – the normal distribution, 2 – the distribution law of the residuals

$$\Pi = \begin{bmatrix} 11.44 \\ 5.092 \\ -26.33 \\ 4.59 \end{bmatrix} \begin{bmatrix} 1,00 & \underset{(0,174)}{-0,005} & \underset{(0,220)}{38,34} & +76,30 & \underset{(60,761)}{-1700,15} \end{bmatrix} = \alpha\beta^{\mathrm{T}} \quad (14)$$

Thus, the vector of the coefficients β in the expression (14) defines relation between the parameters *NOx, Speed,   Load, and   θ:*

$$NOx_t = -0,005 \cdot Speed_t + 38,345 \cdot Load_t + 76,308 \cdot \theta_t - 1700,153 + \varepsilon_t \quad (15)$$

We obtain also the following relations

$$CH_t = -0,004 \cdot Speed_t - 0,612 \cdot Load_t + 2.17 \cdot \theta_t - 3,75 + \varepsilon_t \quad (16)$$

$$Smoke_t = -0,012 \cdot Speed_t + 6,035 \cdot Load_t + 0,227 \cdot \theta_t - 2,65 + \varepsilon_t \quad (17)$$

Estimation errors show the adequacy of models (18) – (20). The correspondent graphs are in Fig. 3.

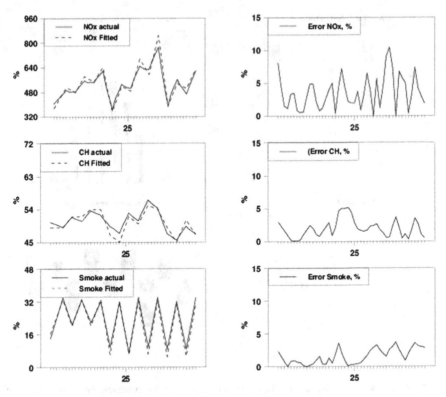

Fig. 3. Analysis of result of modeling

Having assumed that for every functioning mode the values of the speed and load remain constant in the vicinity of the working points, substitution of the Eqs. (5)–(7) in the Eq. (4), allows to reduce the problem of multicriteria optimization to the optimization with single criterion, that is to the choice of the values of parameter θ, when the following criterion is minimum:

$$K(\theta) = (a + \beta_3 \cdot \theta)^{0.5} \cdot (b + \phi_3 \cdot \theta)^{0.3} \cdot (c + \eta_3\theta)^{0.2} \rightarrow \min \qquad (18)$$

As a result of calculation the Table 1 shows the values of the advance angle of fuel injection, which corresponds to the minimal concentration of toxic components in the exhaust gases for the studied diesel.

Table 1. The values of the advance angle of fuel injection for the different diesel mode

Speed	Load, Hм			
	10.00	7,50	6,50	5,50
1500	17.80	19.23	19.80	20.26
2000	17.92	19.31	19.88	20.34
2500	18.00	19.40	19.97	20.43

Table 2. The results

1	2	3	4	5	6	7	8	9	10
Speed	1500			2000			2500		
Parameters	Old	new	Difference, %	Old	new	Difference, %	Old	new	Difference, %
NOx	450	29	93	500	47	90	520	65	88
CH	50	35	30	48	36	25	47	35	12
Smoke	26	25	10	20	22	-10	20	17	15
ge	325	327	-0,6	323	330	2	320	320	0

The results of the use of new values of the advance angle of fuel injection are presented in Table 2.

Table 2 shows that for the diesel with the fixed design parameters at the given characteristics of the fuel equipment (for instance, injection pressure) the proposed approach allows to lower the concentration of toxic components in the diesel exhaust gases without significant increase of the fuel consumption. For example, the maximum increase of the average fuel efficiency compared to the initial value is 2 % (see the column 7 in the Table 2). Moreover, the proposed algorithm of the lowering diesel toxicity, alternatively to the algorithm proposed in [1] allows lowering the concentration of toxic components not only for the steady modes, but also for unsteady modes: during acceleration and deceleration. The advantages of the proposed approach for stationary modes diesel engine at constant load 7.50 Nm are presented in Table 3.

Table 3. The results

Speed	1500			2000			2500		
Parameters	Old method	New method	Difference, %	Old method	New method	Difference, %	Old method	New method	Difference, %
NOx	85	29	65	73	47	36	127	65	49
CH	39	35	10	38	36	5	34	35	-3
Smoke	28	25	10	22	22	0	17	17	0
ge	330	327	0,9	350	330	6	335	320	4

Acknowledgements. The authors express their deep gratitude to the Department of Information Systems of the Sevastopol National Technical University for the support of this research.

References

1. Pervukhina, E., Osipov, K.: Reducing toxicants in the diesel exhaust gas based on optimal fuel injection timing. In: Proceedings Volume of the EURO Mini-Conference on Optimization in the Natural Sciences, Aveiro, Portugal, February 5–9, h.25 (2014)
2. Markov, V.A., Furman, V.V., Mironov, V.A.: Experimental studies of the electronic system of locomotive diesel fuel control. In: Proceedings of Higher Educational Institutions. Machine Building, vol. 1, pp. 38–48 (2012) (in Russian)
3. Wang, X., Stone, C.R.: A study of combustion, instantaneous heat transfer, and emissions in a spark ignition engine during warm-up. In: Proceedings Institution of Mechanical Engineers, vol. 222 Part D, pp. 607–618 (2008)
4. Mollenhauer, K., Tschoeke, H.: Handbook of Diesel Engines, p. 636. Springer, Heidelberg (2010)
5. Salov, T.Y., Tursunov, A.A., Mazhitov, B.J.: Environmental performance evaluation diesel at mountain operation. www.ttu.tj/userfiles/vestnik/vn13.pdf (in Russian)
6. Pervukhina, E., Osipov, K., Rapatski, Y.: Calculating the duration of fuel injection to reduce the concentration of toxic components in the combustion products. Int. Combust. Engines **1**, 80–83 (2013). (in Russian)
7. Orlin, A., Kruglov, M.G.: Internal Combustion Engines: Theory and combined piston engines, p. 372. Mechanical Engineering, Moscow (1983). (in Russian)
8. Pervukhina, E., Osipov, K., Rapatskiy, Y.: Improvement of the acceptance test procedure for the external combustion engine after assembly by using the relationship between diagnostic parameters. J. Mach. Manuf. Reliab. **40**(2), 171–175 (2011). © Allerton Press, Inc. (in Russian)
9. Golikova, V., Pervukhina, E., Sopin, P.: Statistical modeling of machines on diagnostic parameters. J. Mach. Manuf. Reliab. **37**(6), 612–617 (2008). (in Russian)
10. Turner, J.D., Austin, L.: A review of current sensor technologies and applications within automotive and traffic control systems. In: Proceedings of the Institution of Mechanical Engineers, vol. 2014, Part D, pp. 589–614 (2000)
11. Doan, T.A.: RATS software package, User's manual, Version 8.0, Illinois: ESTIMA (2010)

Desirability Functions in Multiresponse Optimization

Başak Akteke-Öztürk[1]([⊠]), Gerhard-Wilhelm Weber[2], and Gülser Köksal[1]

[1] Department of Industrial Engineering, METU, Ankara, Turkey
bozturk@metu.edu.tr
[2] Institute of Applied Mathematics, METU, Ankara, Turkey

Abstract. Desirability functions (DFs) play an increasing role for solving the optimization of process or product quality problems having various quality characteristics to obtain a good compromise between these characteristics. There are many alternative formulations to these functions and solution strategies suggested for handling their weaknesses and improving their strength. Although the DFs of Derringer and Suich are the most popular ones in multiple-response optimization literature, there is a limited number of solution strategies to their optimization which need to be updated with new research results obtained in the area of nonlinear optimization.

1 Introduction

Most industrial processes and products have more than one quality response; they are usually conflicting but should be optimized concurrently and concertedly. For quality improvement, optimal levels of continuous variables (input variables, or factors) are searched which give the best trade-off of these quality characteristics (output variables, or responses). This is a multi-objective optimization problem having a special name i.e., multi-response optimization (MRO). Most commonly used approaches to solve the multi-response (surface) optimization problems utilize *response surface methodology (RSM) Taguchi method, loss functions, Mahalanobis distance,* and *desirability function approach* (Khuri 1996; Logothetis and Wynn 1989; Miettinen 1999; Montgomery 2000). Each of these approaches has its own limitations.

In multi-response (surface) optimization problems, statistical design of experiments is commonly used to collect data (Montgomery 2000). In an experimental design, the first step consists of defining the problem. Then, the factors together with their ranges and specific levels are chosen at which experimental runs will be made. The next step is the estimation of the response models that relate the factors to the responses. Finally, an appropriate experimental design layout is designed which is robust to different sources of variability in the data so that the most desirable process and product parameter settings are found. The most common way of obtaining the response models is *regression* by means of *polynomial fitting* or *spline fitting.* For the cases where polynomial fitting is not capable

© Springer International Publishing Switzerland 2015
A. Plakhov et al. (Eds.): EmC-ONS 2014, CCIS 499, pp. 129–146, 2015.
DOI: 10.1007/978-3-319-20352-2_9

of modeling the quantitative and qualitative responses, *artificial neural networks* also used in the literature.

1.1   Desirability Function (DF) Approach

Optimization of multiple responses at the same time has the challenge that each of them may have a different measurement scale. When we use DF approach, a scale-free value between 0 and 1 is assigned to all responses in the problem by the so-called individual DFs, and then combine these values by taking usually their arithmetic or geometric mean yielding a single objective to obtain an overall DF having a value in the interval $[0, 1]$. There are different versions DFs in the literature. However, many software programs such as Minitab, JMP, Package Desirability and Design-Expert that are employed in industrial applications adopt DFs of Derringer and Suich type in their multi-response optimization support.

The philosophy behind the DF approach is that when one of the quality characteristics of an industrial process or product with many characteristics is not in the desired limits, then the overall quality is not desirable. DF approach has originally been introduced by Harrington (1965) with exponential individual DFs which are aggregated by geometric mean to obtain the overall DF. Derringer and Suich (1980) has proposed the version of individual DFs described in Eqs. (2) and (3) below, offering more shape alternatives. DFs of Derringer and Suich (1980) may be nondifferentiable at a target value between the lower and upper acceptable bounds of the response based on the values of the shape parameter of the function. Derringer (1994) develops the weighted case of these DFs as an improvement of the approach for assigning relative importance (or priorities) to the individual responses. The approach of Ch'ng et al. (2005) is based on the arithmetic mean aggregation of the proposed linear individual DFs that are continuously differentiable everywhere within their domains by definition and having desirability in the interval $[0, 2]$. In Wu and Hamada (2000), so-called double-exponential DFs are proposed to avoid the difficulty of choosing proper lower and upper bound values of responses of Derringer and Suich formulation of DFs. An interactive DFs approach is introduced recently in Jeong and Kim (2008) which takes into account the preference of the decision maker(s) on the trade-offs among the responses or on the shape, bound and target of a DF.

1.2   Correlated Responses in DFs

Kim and Lin (2000) suggests a MRO approach which uses DFs of exponential structure for maximizing the degree of overall satisfaction with respect to all the responses and show that their approach is robust to possible dependencies between responses. Although possible statistical and preferential dependencies of responses are usually ignored in DF approach Fuller and Scherer (1998), there are some improvements in considering possible correlations between responses and variances of them. Chen et al. (2012) augments the DFs of Harrington to add this capability by minimizing the variances of the predicted responses.

Double-exponential DFs are also improved in this respect in Wu (2004). Double-exponential DFs based on Taguchi's loss function are redefined to extend the DF model for correlated multiple quality characteristics. The approach presented in Wu (2009) takes into account correlated responses by inserting correlated desirability value into the overall DF formulation. The literature on DFs continues to extend with different modifications in the formulations, aggregation techniques and optimization criteria. A detailed review of DFs is also provided in Murphy et al. (2005) and Fogliatto (1998).

1.3   Variance Information in MRO

Robust design (RD) of multiple responses focus on developing a fitted-response (surface) model that accurately reflects the true variability of a system (static or dynamic), noise factors (uncontrollable variables) and the appropriate quality characteristics (or responses) of interest. The uncertainty associated with the fitted-response (surface) model is known as response model uncertainty. There are two aspects related with this: responses' models differ in terms of the quality of predictions (variance due to uncertainty in the regression coefficients i.e., a response model predicts/performs better) or responses' models are characterized by unequal sensitivity to uncontrollable variables (robustness i.e., a response model is insensitive to noise) (Costa and Loureno 2011; Kovach and Cho 2008).

2   DFs of Derringer and Such Type

In a multi-response optimization problem, response $Y(\mathbf{x})$ is a function $Y : \mathbb{R}^n \rightarrow \mathbb{R}$ of vector of independent variables $\mathbf{x} = (x_1, x_2, \ldots, x_n)^T$, where $x_i \in \mathbb{R}$ $(i = 1, 2, \ldots, n)$. An individual DF $d(Y(\mathbf{x}))$ scales a response into the interval $[0, 1]$, i.e., $d : \mathbb{R} \rightarrow [0, 1]$. This means that the function d becomes 0 for completely undesirable values of the response, and it becomes 1 for totally desirable values of it. We notice that DFs are composite functions of response functions which depend on independent variables (or factors). In this study, we denote an individual DF by d as a function of y and d^Y as a function of $\mathbf{x}$.

$$d^Y(\mathbf{x}) := d(Y(\mathbf{x})) = d(y), \tag{1}$$

where $y := Y(\mathbf{x})$ with $y \in \mathbb{R}$ and $d^Y : \mathbb{R}^n \rightarrow [0, 1]$. There are two types of these functions, *one-sided* and *two-sided* ones according to Derringer and Such (1980):

$$d(y) := \begin{cases} 0, & \text{if } y \le l, \\ \left(\frac{y-l}{u-l}\right)^r, & \text{if } l < y \le u, \\ 1, & \text{if } y > u, \end{cases} \tag{2} \qquad d(y) := \begin{cases} 0, & \text{if } y \le l, \\ \left(\frac{y-l}{t-l}\right)^{s_1}, & \text{if } l < y \le t, \\ \left(\frac{y-u}{t-u}\right)^{s_2}, & \text{if } t < y \le u, \\ 0, & \text{if } y > u. \end{cases} \tag{3}$$

Here, l is the minimum and u is the maximum acceptable value of y, and t is the most desirable value of y. The value of r used in Eq. (2) should be decided by the user. The larger the r, the more desirable are the y values closer to

u, and vice versa. We note that s_1 and s_2 in Eq. (3) have a similar meaning to r. There can be three different types of optimization task for a response: smaller-the-better, larger-the-better or nominal-the-best (see Fig. 1). Definition of individual DFs for the larger-the-better (one-sided DF) case responses is given in Eq. (2). A similar definition can be given for the smaller-the better (one-sided DF) case. Equation (3) corresponds to the DFs of nominal-the best (two-sided DF) type responses. DFs are flexible in the sense that a wide variety of shapes and asymmetric specifications are possible according to the decisions of a decision maker (usually problem owners and experts) of the problem (see Fig. 2).

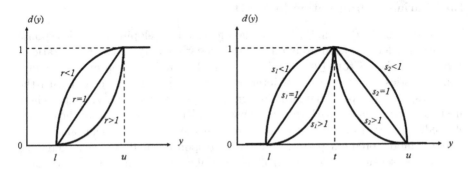

Fig. 1. One-sided and two-sided individual DFs of Derringer and Suich's type.

Assessing desirability of a response means to choose a suitable piecewise-smooth function $d(y)$ ($y = Y(\mathbf{x})$) by deciding the bounds and targets of the DF. Usually most simplistic functions, i.e., piecewise-linear functions including a single nondifferentiable point are preferred. In general, a one-sided individual DF can be either linear or nonlinear but smooth and monotone (i.e., convex or concave) on its domain. A two-sided individual DF can be either piecewise-linear or nonlinear which has at least one nondifferentiable point occuring at its target. In fact, these functions belong to an abstract class of piecewise-smooth functions, i.e., *min-type* functions.

We assume that there are m many responses in a multi-response optimization problem. After calculating the desirabilities of all responses by corresponding functions given in (2) and (3), overall DF $D(\mathbf{y}) : \mathbb{R}^m \to [0,1]$ is calculated using the geometric mean (Derringer and Suich 1980):

$$D(\mathbf{y}) := (d_1(y_1) \cdot d_2(y_2) \cdot \ldots \cdot d_m(y_m))^{\frac{1}{m}}, \tag{4}$$

where $\mathbf{y} := \mathbf{Y}(\mathbf{x})$ and $\mathbf{Y}(\cdot) := (Y_1, Y_2, \ldots, Y_m)^T(\cdot)$. Here, it is obvious that $D(\mathbf{y})$ will have a value in $[0,1]$. We denote the overall desirability as a function of $\mathbf{x}$ by $D^Y : \mathbb{R}^n \to [0,1]$ and define it by $D^Y(\mathbf{x}) := D(\mathbf{Y}(\mathbf{x}))$, i.e., $D^Y(\mathbf{x}) := (d_1^Y(\mathbf{x}) \cdot d_2^Y(\mathbf{x}) \cdot \ldots \cdot d_m^Y(\mathbf{x}))^{\frac{1}{m}}$. The function given in (4) is a nonlinear composite objective function including signomial terms.

Employing geometric mean to compute the overall desirability from individual desirabilities gives rise to the main property of DFs as an approach. If a

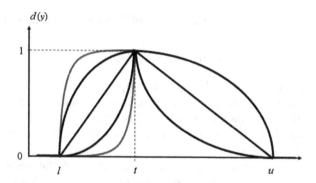

Fig. 2. Asymmetric individual DF

desirability d of a response y becomes 0 at a factor value $\bar{x}$, the overall desirability becomes 0 at this $\bar{x}$, independently from the values of other individual desirabilities at that point. In this formulation of DFs, possible correlations between the responses are not taken into account and hence, it is assumed that the responses are independent of each other.

When the importance of individual DFs may differ in computing the overall DFs, a weighting strategy is possible (Derringer 1994):

$$D(\mathbf{y}) := \left(\prod_{j=1}^{m} d_j(y_j)^{w_j} \right)^{\frac{1}{\sum_{j=1}^{m} w_j}}. \tag{5}$$

Weighted overall desirability has similar properties to the non-weighted one. Again, if one of the responses is undesirable at a factor vector $\bar{x}$, then the overall desirability is zero at that point, i.e., $D(\mathbf{Y}(\bar{x})) = 0$, without considering desirabilities of other responses at that point. These weights can be specified by the decision maker next to the shapes of the curves of DFs. Obviously, when deciding about the weights it would be better to take into account relative importance of the product and process responses with respect to each other.

2.1 Optimization of Overall DF

The overall DF $D(\mathbf{y})$ is a continuous function of the individual desirabilities $d_j(y)$ from Eq. (4) and we see that each function d is continuous up to y from (2) and (3). In this study, a response Y is assumed to be a continuous function of the vector of factors, $\mathbf{x}$. Therefore, the overall DF D^Y is a continuous function of the factor vector $\mathbf{x}$.

Herewith, the problem takes the following form:

maximize $D(\mathbf{Y}(\mathbf{x}))$
subject to
 i. bounds of the factors x_i $(i = 1, 2, \ldots, n)$,
 ii. bounds and targets of the responses $Y_j(\mathbf{x})$ $(j = 1, 2, \ldots, m)$,
$$\tag{6}$$

Here, $\mathbf{x} = (x_1, x_2, \ldots, x_n)^T$ is the vector of factors. The constraint set is a parallelepiped

$$[\mathbf{l}_x, \mathbf{u}_x] := \underset{i=1}{\overset{n}{\times}} [l_{x_i}, u_{x_i}], \tag{7}$$

where l_{x_i} is the lower limit and u_{x_i} is the upper limit of x_i ($i = 1, 2, \ldots, n$). First constraint group of the problem (6), i.e., the bounds for the factors, are decided during the experimental design, and hence, they are known during the optimization procedure. In computation, the bounds for the factor levels are usually standardized to $[-1, 1]$.

Second constraint group, i.e., bounds and targets of the responses $\mathbf{Y} = (Y_1, Y_2, \ldots, Y_m)^T$ are decided during experimental design and don't change. We assume a static system in this study. These bounds l_j and u_j for $Y_j(\mathbf{x})$ are in fact functions of $\mathbf{x}$, usually as 95% confidence interval. In computations, non-negativity constraints for the individual DFs are added to prevent the algorithm from stopping at values quite near to 0 from below.

Problem (6) can be expressed implicitly by combining bound constraints:

$$\text{maximize } D^{\mathbf{Y}}(\mathbf{x}) = \left(d_1^Y(\mathbf{x})\right)^{w_1} \cdot \left(d_2^Y(\mathbf{x})\right)^{w_2} \cdot \ldots \cdot \left(d_m^Y(\mathbf{x})\right)^{w_m} \tag{8}$$
$$\text{subject to } \mathbf{x} \in \mathbb{X} \cap \mathbb{I}^X.$$

Here, we assume that $w_1 + w_2 + \ldots + w_m = 1$ without loss of generality not all zero at the same time ($j = 1, 2, \ldots, m$). If all weights were zero at same time, then any $y_j \in \mathbb{R}$ would become a solution of problem (8). We notice that any sum of weights, say $\omega_1 + \omega_2 + \ldots + \omega_m = r$ for some $r > 0$, can be reduced to 1 by defining $w_j := \omega_j / r$ ($j = 1, 2, \ldots, m$).

We have $\mathbf{x} \in \mathbb{X} \subset \mathbb{R}^n$ and $x_i \in \mathbb{X}_i$ where $\mathbb{X}$ is the Cartesian product of regions $\mathbb{X}_i \subset \mathbb{R}$ ($i \in I = \{1, 2, \ldots, n\}$):

$$\mathbb{X} = \underset{i \in I}{\times} \mathbb{X}_i \quad (= \mathbb{X}_1 \times \mathbb{X}_2 \times \ldots \times \mathbb{X}_n). \tag{9}$$

We remember that every response $y_j = Y_j(\mathbf{x})$ is desired in some interval $\mathbb{I}_j = [l_j, u_j]$ ($j = 1, 2 \ldots, m$). Now, we define

$$\mathbb{I}^X := \{\mathbf{x} \in \mathbb{R}^n \mid Y_j(\mathbf{x}) \in \mathbb{I}_j \ (j = 1, 2, \ldots, m)\} = \bigcap_{j=1}^{m} \left(Y^{-1}([l_j, u_j])\right). \tag{10}$$

Here, $\mathbb{I}^X$ is closed as it is the finite intersection of the closed sets $Y^{-1}([l_j, u_j]) \subset \mathbb{R}^n$. Similarly, $\mathbb{X}$ is closed. Hence, $\mathbb{X} \cap \mathbb{I}^X$ is compact and $D^Y(\mathbf{x})$, the objective function of (8), is continuous. We conclude that a globally optimal point to problem (8) always exists, but it may not be unique (Ozdaglar and Tseng 2006).

Optimization of overall DF given in Eq. (4) using gradient-based approaches becomes a complicated task when there are nondifferentiable two-sided individual DFs in the problem. In the two-sided DFs formulation (3), the target value may be attained at a nondifferentiable point, and hence, the function is not smooth at this point. It follows that a suitable single objective optimization method shall be chosen to solve the optimization problem of maximizing the continuous but nondifferentiable overall DF, i.e., we want $D(\mathbf{y})$ as close to unity as possible. In Derringer and Such (1980), a univariate search technique that do not employ derivative information is implemented in FORTRAN on the overall desirability values evaluated at all factor levels of the response surface design to solve the overall optimization problem. Other derivative free optimization techniques suitable for this problem are pattern search, direct search methods (Derringer and Such 1980) and mesh-adaptive direct search methods (Conn et al. 2009). Castillo et al. (1996) demonstrates a modification of piecewise-linear individual DF of Derringer and Such as a function of a response. This modification is based on a polynomial approximation to smoothen individual DFs at their nondifferentiable points. Hence, the optimization problem of the DF becomes a nonlinear continuously differentiable problem and is solved by the generalized reduced gradient method (GRG2 of MS Excel) (Lasdon et al. 1978). It is possible to apply other gradient-based methods to this modified DFs.

In Akteke-Öztürk et al. (2014), we show how to use nonsmooth and global optimization approaches. We suggest some transformation strategies in combination with a reformulation of individual DFs. Our main software environment is GAMS and its solvers (CONOPT and BARON). Nonsmooth optimization approaches continue to develop in generalizing notions of differential optimization such as gradient, convexity, and Lagrangian to solve problems including nondifferentiable functions (Clarke 1983; Demyanov and Rubinov 1986; Dutta 2005; Gasimov and Ustun 2007; Lemarechal 1978). Global optimization approaches solve nonlinear optimization problems to global optima in case of existency of global solution(s) (Pardalos and Romeijn 2002; Gershon and Shaked 2012). One of our strategies is to use MSG algorithm to obtain sharp augmented Lagrangian of the problem. Other strategy is to obtain the continuous relaxation of the problem which is originally a signomial geometric programming problem (Ryoo and Sahinidis 1996; Tawarmalani 2002; Gershon and Shaked 2012).

3   Additive Overall DF

A two-sided individual desirability function (DF) can be expressed as:

$$d(y) = \begin{cases} 0, & \text{if } y \le l, \\ d_1(y), & \text{if } l < y \le t, \\ d_2(y), & \text{if } t < y \le u, \\ 0, & \text{if } y > u, \end{cases} \quad \text{where} \quad d_1(y) := \left(\frac{y-l}{t-l}\right)^{s_1} \quad \text{and} \quad d_2(y) := \left(\frac{y-u}{t-u}\right)^{s_2}.$$

$$(11)$$

Here, l is the minimum and u is the maximum acceptable value of y, and t is the most desirable value of y. The value of s_1 and s_2 should be specified by the user. We can express each individual function d with a mixed-integer formulation by taking a convex combination of sides d_1 and d_2, using the binary integer variable $z \ (= z_j) \ (j = 1, 2, \ldots, p)$, $z = 0, 1$:

$$d(y, z) = z d_1(y) + (1 - z) d_2(y).$$

$$(12)$$

Here, the binary coefficient z becomes 1 when $d_1(y)$ is active (on). By using Eqs. (1) and (12), we can reach the mixed-integer formulation in $\mathbf{x}$ of a two-sided individual DF:

$$d^Y(\mathbf{x}, z) = z d_1^Y(\mathbf{x}) + (1 - z) d_2^Y(\mathbf{x}). \tag{13}$$

Let us assume that there are m many responses in a multi-response optimization problem, we denote by p the number of responses having two-sided desirabilities and by $m - p$ the number of responses having one-sided DFs, where $0 < p \le m$. We can define adjusted overall DF $D^Y(\cdot)$ as follows:

$$D^Y(\mathbf{x}, \mathbf{z}) := \left[\prod_{j=1}^{p} d_j^Y(\mathbf{x}, z_j)^{w_j} \cdot \prod_{j=p+1}^{m} d_j^Y(\mathbf{x})^{w_j} \right]^{\frac{1}{\sum_{j=1}^{m} w_j}}, \tag{14}$$

where $w_j \ge 0$ $(j = 1, 2, \ldots, m)$, $\mathbf{x} := (x_1, x_2, \ldots, x_n)^T$ and $\mathbf{z} := (z_1, z_2, \ldots, z_p)^T$. The optimization problem of the adjusted overall DF can be stated as follows:

$$(\mathcal{P}) \quad \begin{cases} \text{maximize } D^Y(\mathbf{x}, \mathbf{z}) \\ \text{subject to } \mathbf{x} \in \mathbb{X} \cap \mathbb{I}^X, \\ \qquad\qquad \mathbf{h}(\mathbf{z}) = \mathbf{0}_p. \end{cases} \tag{15}$$

We apply negative logarithm to the overall DF from Eq. (14), i.e., $F(\mathbf{Y}(\cdot)) := -\log(D(\mathbf{Y}(\cdot)))$ and obtain an additive expression with respect to the individual desirabilities $f_j(Y_j(\cdot)) = -\log d_j$:

$$F(\mathbf{Y}(\cdot)) := \sum_{j=1}^{m} w_j f_j(Y_j(\cdot)), \tag{16}$$

for $w_j \ge 0$ $(j = 1, 2, \ldots, m)$, and not all being zero at the same time $(j = 1, 2, \ldots, m)$. We may assume that $w_1 + w_2 + \ldots + w_m = 1$ without loss of generality. If all weights were zero at same time, i.e., $w_j = 0$ $(j = 1, 2, \ldots, m)$, then any $y_j \in \mathbb{R}$ would become a solution of the problem $(\mathcal{P})$. Any sum of weights, say $w_1 + w_2 + \ldots + w_m = r$ for some $r > 0$, can be reduced to 1 by defining $w_j := w_j/r$ $(j = 1, 2, \ldots, m)$. The additive (separable) overall DF $F^Y(\cdot) = F(\mathbf{Y}(\cdot)) = F(\mathbf{y})$, where $F : \mathbb{R}^m \to \mathbb{R}$ and $F^Y : \mathbb{R}^n \to \mathbb{R}$ is the weighted sum of individual $f_j(Y_j(\mathbf{x}))$ $(j = 1, 2, \ldots, m)$. The individual functions $f_j : \mathbb{R} \to \mathbb{R}$ $(j = 1, 2, \ldots, m)$ are again composite functions and always positive. Since d_j becomes zero at the lower and upper bound of a response, we can consider applying following techniques to make sure that the logarithm becomes always defined.

Cutting-off the Interval of y (*Technique 1*): We introduce lower bounds $\delta^l > 0$ and $\delta^u > 0$ (arbitrarily small numbers) for $Y(\mathbf{x}) - l$ and $u - Y(\mathbf{x})$, respectively. This can be interpreted as cutting-off of a piece of the interval of $Y(\mathbf{x})$ at l of length δ^l and at u of length δ^u. By cutting-off the half neighborhoods $[l, l + \delta^l)$ and $(u - \delta^u, u]$, which are mapped into the intervals $[0, \epsilon^l]$ and $[0, \epsilon^u]$

($\epsilon^l := d(l + \delta^l)$ and $\epsilon^u := d(u - \delta^u)$ are arbitrarily small numbers), respectively, the function is prevented from entering these intervals and desirabilities never become zero as shown in Fig. 3.

The new DF will be defined on the interval $[l + \delta^l, u - \delta^u]$ and we always have $d(Y(\mathbf{x})) > 0$, whereas it will not be defined on the intervals $[l, l + \delta^l)$ and $(u - \delta^u, u]$. We may add two more constraints to the optimization problem from $(\mathcal{P})$ which will not affect the solution but ensure that the two-sided desirabilities never vanish for all $j = 1, 2, \ldots, m$:

$$(Y(\mathbf{x}) - l)^2 \geq (\delta^l)^2, \quad \text{and}$$
$$(u - Y(\mathbf{x}))^2 \geq (\delta^u)^2. \tag{17}$$

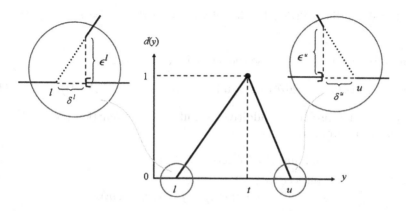

Fig. 3. Cutting-off of individual desirability functions $d_j(y_j)$ $(j = 1, 2, \ldots, m)$.

Shifting the DFs (*Technique 2*): We introduce ϵ-individual two-sided DFs with $d^\epsilon(Y(\mathbf{x})) := (d + \epsilon)(Y(\mathbf{x}))$, where $\epsilon := \epsilon_{Y(\mathbf{x})} > 0$ (arbitrarily small number) to be a lower bound for ϵ-individual desirabilities, $d^\epsilon(Y(\mathbf{x})) \geq \epsilon$. This can be interpreted as a shift in the function values from $[0, 1]$ to $[\epsilon, 1 + \epsilon]$. Hence, at $Y(\mathbf{x}) = l$ and $Y(\mathbf{x}) = u$, we prevent DF being zero as shown in Fig. 4.

By doing this, we change the definition of the individual DFs with respect to ϵ as follows:

$$d^\epsilon(y) := \begin{cases} \epsilon, & \text{if } y \leq l, \\ (\frac{y-l}{t-l})^{s_1} + \epsilon, & \text{if } l < y \leq t, \\ (\frac{y-u}{t-u})^{s_2} + \epsilon, & \text{if } t < y \leq u, \\ \epsilon, & \text{if } y > u. \end{cases} \tag{18}$$

Remark 1. Our Techniques 1 and 2 are suitable for both linear and nonlinear versions of individual DFs. The small numbers δ^l, δ^u, ϵ^l, ϵ^u and ϵ should be chosen according to this shape of the DF near l and u. Moreover, we note that our original problem is a maximization one optimal solution of which will not be affected by this cutting-off and shifting.

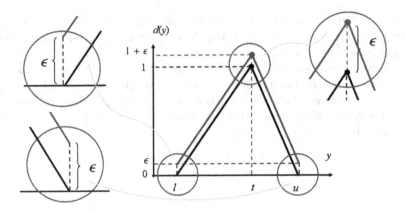

Fig. 4. Shifting individual desirability functions $d_j(y_j)$ $(j = 1, 2, \ldots, m)$.

Among those techniques, we use the shifted DFs $(d^\epsilon(y))$:

$$f(y) := -\log(d^\epsilon(y)) \quad \text{and} \quad f(Y(\cdot)) = -\log(d^\epsilon(Y(\cdot))), \tag{19}$$

The optimization problem of additive overall objective function given in (16) is a minimization problem:

$$\begin{aligned}
\text{minimize} \ & F(\mathbf{Y}(\mathbf{x})) \\
\mathbf{x} \\
\text{subject to} \ & \mathbf{x} \in \mathbb{X}, \\
& Y_j(\mathbf{x}) \in \mathbb{I}_j \quad (j = 1, 2, \ldots, m).
\end{aligned} \tag{20}$$

Here, $\mathbf{x} \in \mathbb{X} \subset \mathbb{R}^n$, where the parallelepiped $\mathbb{X}$ given in Eq. (9) is implicitly defined by a finite number of inequality constraints and $Y_j(\mathbf{x}) \in \mathbb{I}_j \subseteq \mathbb{R}$, where $\mathbb{I}_j$ is an interval.

Remark 2. We note that this reformulation of the overall problem does not cause any change in the global optimal solution, i.e., the solution of the original problem $(\mathcal{P})$ of Eq. 15 and the one of Eq. (20) are the same. However, we could not say the same thing for the solution if $F(\mathbf{Y}(\mathbf{x}))$ were the weighted sum of the individual DFs $d_j(Y_j(\mathbf{x}))$. For a review of different reformulations similar to the DF optimization and their solution characteristics, we refer to the study of Conn et al. (2009).

3.1   A Finite Number of Nondifferentiable Points in DFs

Although in practice two-sided DFs typically include only one nondifferentiable point each, there can be more than one nondifferentiable points in a two-sided individual DF reflecting the desired behavior of the function around its target point. Individual DFs including a finite number of nondifferentiable point are especially useful for approximation of nonlinear conventional DFs by linear or affine functions.

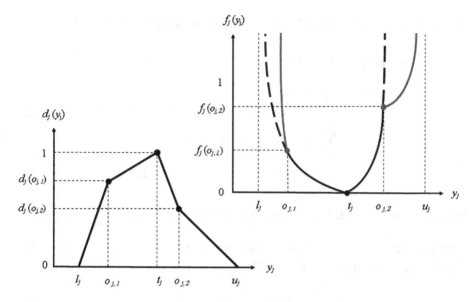

Fig. 5. An individual DF with 3 nondifferentiable points $d_j(y_j)$ and $f_j(y_j)$ after negative logarithm is applied.

Regarding its shape, we see that the DF shown in Fig. 4. has a similar tendency to target value as in $s_1 < 1$ and $s_2 > 1$ case of a conventional two-sided individual DF defined in Eq. (11). Let us assume that there are $\zeta_j - 1$ ($\zeta_j \in \mathbb{R}$) ($j = 1, 2, \ldots, m$) many nondifferentiable points and hence, ζ_j "pieces" in individual DFs of a multi-response optimization problem. Here, we name the part of the function lying between consecutive nondifferentiable points by the word "piece". A "side" is the part of the function between lower and target values of the response where in between there can be a finite number of nondifferentiable points. In DFs including a single nondifferentiable point the meanings of these notions coincide. In Fig. 5, an example of individual DF having 2 sides but 4 pieces because of 3 nondifferentiable points and its possible outcome with applying negative logarithm is shown. A response value corresponds to a unique combination of the factor values which means one 'piece' of the function becomes active (on) while the remaining ones are inactive (off). This makes introduction of adjusted DFs given in Eq. (12) to the related optimization problem a suitable approach. An individual DF $f_j^Y(\mathbf{x}, \mathbf{z}_j)$ including $\zeta_j - 1$ many nondifferentiable points can be expressed as follows for $\mathbf{z}_j = (z_{j,1}, z_{j,2}, \ldots, z_j^{\zeta_j})^T$ with

$$\sum_{\zeta=1}^{\zeta_j} z_j^\zeta = 1 \ (j = 1, 2, \ldots, m) \text{ and } z_j^\zeta \in \{0, 1\} \ (\zeta = 1, 2, \ldots, \zeta_j; j = 1, 2, \ldots, m):$$

$$f_j^Y(\mathbf{x}, \mathbf{z}_j) = \sum_{\zeta=1}^{\zeta_j} z_j^\zeta f_j^{Y,\zeta}(\mathbf{x}) \quad (j = 1, 2, \ldots, m). \tag{21}$$

Hence, the overall problem $(\mathcal{P})$ turns into a minimization problem:

$$
\begin{aligned}
\text{minimize} \quad & \sum_{j=1}^{m} w_j f_j^Y(\mathbf{x}, \mathbf{z}_j) \\
\text{subject to} \quad & \mathbf{x} \in [\mathbf{l}_x, \mathbf{u}_x], \\
& f_{j}^{Y,\zeta}(\mathbf{x}) \geq 0 \quad (\zeta = 1, 2 \ldots \zeta_j; j = 1, 2, \ldots, m), \\
& \sum_{\zeta=1}^{\zeta_j} z_j^\zeta = 1 \quad (j = 1, 2, \ldots, m), \\
& z_j^\zeta \in \{0, 1\} \quad (\zeta = 1, 2, \ldots, \zeta_j; j = 1, 2, \ldots, m).
\end{aligned}
\tag{22}
$$

where $w_j \geq 0$ $(j = 1, 2, \ldots, m)$ are given weights and $\sum_{j=1}^{m} w_j = 1$. We note that each function $f_j^{Y,\zeta} : \mathbb{R}^n \to \mathbb{R}$ is assumed to be a C^2-function. Here, z_j^ζ is the indicator of the active piece $f_j^{Y,\zeta}$ of f_j^Y:

$$
z_j^\zeta = \begin{cases} 1, \text{if} \quad f_j^Y(\mathbf{x}) = f_j^{Y,\zeta}(\mathbf{x}), \\ 0, \text{otherwise.} \end{cases}
\tag{23}
$$

The constraint $z_j^\zeta \in \{0, 1\}$ can equivalently be stated as $z_j^\zeta - (z_j^\zeta)^2 = 0$ and $\mathbf{z}_j = (z_{j,1}, z_{j,2}, \ldots, z_j^{\zeta_j})^T$ is a unit vector of length ζ_j. Problem (22) is a nonconvex global optimization problem of a nonsmooth objective function with possibly many local minima and maxima. By using max-type functions, we can state a special case of problem (22) as follows:

$$
\begin{aligned}
\text{minimize} \quad & F^Y(\mathbf{x}) \quad \left(= \sum_{j=1}^{m} w_j \max_{\zeta=1,2,\ldots,\zeta_j} f_j^{Y,\zeta}(\mathbf{x})\right) \\
\text{subject to} \quad & \mathbf{x} \in [\mathbf{l}_x, \mathbf{u}_x], \\
& F^Y(\mathbf{x}) \geq \mathbf{0},
\end{aligned}
\tag{24}
$$

where $\mathbf{x} = (x_1, x_2, \ldots, x_m)^T$ and the objective function is a convex combination of the max-type functions $f_j^{Y,\zeta}$.

3.2   Separation of Parameters as y-Space and x-Space

The overall function $F(\cdot)$ of the vector $\mathbf{y} = (y_1, y_2, \ldots, y_m)^T$, where $\mathbf{y} = \mathbf{Y}(\mathbf{x})$ satisfies $F(\mathbf{y}) := \sum_{j=1}^{m} w_j f_j(y_j)$ for $f_j : \mathbb{I}_j \to \mathbb{R}$ $(j = 1, 2 \ldots, m)$. The intervals $\mathbb{I}_j := [l_j, u_j]$ $(j = 1, 2 \ldots, m)$ can be disjoint for all j, intersecting or even the same for some or all j. We notice that the graph of F in $\mathbf{y}$ can be connected or disconnected according to the positions of these intervals; usually it is disconnected, and it can be connected only if all responses have the same interval for all j. If we consider the optimization of F with respect to $\mathbf{y}$, we have the following constrained minimization problem:

$$\underset{\mathbf{y}}{\text{minimize}}\ F(\mathbf{y}) = w_1 f_1(y_1) + w_2 f_2(y_2) + \ldots + w_m f_m(y_m)$$
$$\text{subject to } y_j \in \mathbb{I}_j \quad (j = 1, 2, \ldots, m), \tag{25}$$

where $w_1 + w_2 + \ldots + w_m = 1$ with $w_j \geq 0$ $(j = 1, 2, \ldots, m)$ representing the convex combinations of the individual DFs f. By the additive structure of the objective function, the optimal (global) solution of this problem $\mathbf{t}$ is a vector of the (global) optimal solutions of $f_j(y)$, say t_j $(j \in J)$: $\mathbf{t} := (t_1, t_2, \ldots, t_m)^T$, i.e., the vector of what we called the *target point*. This vector $\mathbf{t}$ is usually named as the *ideal point* of the overall problem in Eq. (22); it lies in the m-dimensional cube $\mathbb{I} \subset \mathbb{R}^m$:

$$\mathbb{I} = \underset{j \in J}{\times}\ \mathbb{I}_j \quad (= \mathbb{I}_1 \times \mathbb{I}_2 \times \ldots \times \mathbb{I}_m). \tag{26}$$

3.3   Structural Configuration of DF

It is possible to have an equivalent formulation of Eq. (22) by using piece-wise max-type functions to express individual functions f_j over their interval $\mathbb{I}_j := [l_j, u_j]$ partitioned suitably into a κ_j number of subintervals $\mathbb{I}_{j,\kappa}$ $(\kappa = 1, 2, \ldots, \kappa_j)$. We mean by a "suitable" partitioning of the interval that at each subinterval corresponding function is of max-type (Fig. 6).

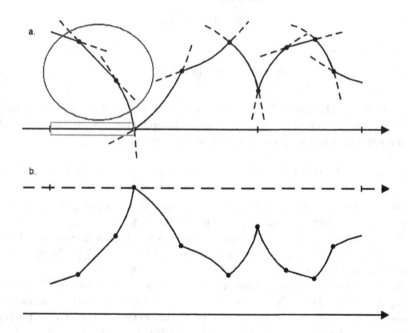

Fig. 6. Min-type and max-type piecewise-differentiable functions

It is obvious that the number of subintervals κ_j could be less than or equal to ζ_j. Let us define the index sets that are used throughout this chapter:

- for the number of individual functions f_j and intervals $\mathbb{I}_j$:
  $J = \{1, 2 \ldots, m\}$ with elements $j \in J$,
- for the number of subintervals of $\mathbb{I}_j$:
  $K_j = \{1, 2, \ldots, \kappa_j\}$ with elements $\kappa \in K_j$,
- for the number of function pieces at each subinterval κ:
  $Z_{j,\kappa} = \{1, 2, \ldots, \zeta_{j,\kappa}\}$ with elements $\zeta \in Z_{j,\kappa}$,
- for the total number of function pieces for each j:
  $Z_j = \{1, 2, \ldots, \zeta_j\}$ with elements $\varsigma \in Z_j$.

Here, $\mathbb{I}_j = \bigcup_{\kappa=1}^{\kappa_j} \mathbb{I}_{j,\kappa}$ where $\mathbb{I}_{j,\kappa} := [l_{j,\kappa}, u_{j,\kappa}]$ is the interval with lower bound $l_{j,\kappa}$ and upper bound $u_{j,\kappa}$ ($\kappa \in K_j, j \in J$). Furthermore, we assume that neighboring subintervals have just boundary points in common: $u_{j,\kappa} = l_{j,\kappa-1}$ ($\kappa \in K_j \setminus \{1\}, j \in J$). At each subinterval $\mathbb{I}_{j,\kappa}$, the max-type function is called $f_{j,\kappa}$, i.e.,

$$f_{j,\kappa} := f_j \mid_{\mathbb{I}_{j,\kappa}} \quad (\kappa \in K_j, j \in J) \text{ where } \quad f_{j,\kappa} := \max_{\zeta=1,2,\ldots,\zeta_j} f_{j,\kappa}^\zeta. \quad (27)$$

The minimization of functions $f_{j,\kappa}$ in y for a fixed $\kappa \in K_j$ and for a fixed $j \in J$ is a finitely constrained nonsmooth minimax problem:

$$\begin{aligned} & \underset{y}{\text{minimize }} f_{j,\kappa}(y) \\ & \text{subject to } y \in \mathbb{I}_{j,\kappa}, \end{aligned} \quad (28)$$

and, in other words,

$$\underset{y \in \mathbb{I}_{j,\kappa}}{\text{minimize }} \underset{\zeta \in Z_{j,\kappa}}{\text{maximize }} f_{j,\kappa}^\zeta(y). \quad (29)$$

Let the solution of (28) be called $t_{j,\kappa}$ ($\kappa \in K_j$) for each $j \in J$. Now, the minimization of individual functions $f_j(y)$ is a *discrete* optimization problem, actually, an *enumeration problem*, over all κ, for each regarded j:

$$\min_{y \in \mathbb{I}_j} f_j(y) := \min\{f_{j,1}(t_{j,1}), f_{j,2}(t_{j,2}), \ldots, f_{j,\kappa_j}(t_{j,\kappa_j})\} = \min_{\kappa \in K_j} f_{j,\kappa}(t_{j,\kappa}). \quad (30)$$

By solving these problems to find the minimum of $f_j(y)$ per given $j \in J$, we obtain a set of solutions, say t_j, which are, in fact, $t_j := t_{j,\bar{\kappa}_j}$ at a certain $\bar{\kappa}_j \in K_j$. These solutions are the target points as we discuss in Sect. 3.2. Hence, the vector $\mathbf{t} := (t_1, t_2, \ldots, t_m)^T$ is the *ideal point* of the overall problem that lies in the m-dimensional cube $\mathbb{I}$.

The piecewise smooth structure of the functions $f_j(\cdot)$ together with the additive separability (also called linearity) of the regarded overall function $F(\cdot)$ enables us to be concerned about the local properties of the max-type functions $f_{j,\kappa}(\cdot)$ and their optimization problem (28), to achieve results and gain qualitative insights into the full-dimensional problem (25) in $\mathbf{y}$.

4   A New Approach for Multi-objective Optimization: Two-Stage (Bilevel) Method

If we consider the optimization problem (25) (in $\mathbf{y}$ only), its solution will be the ideal solution $\mathbf{t} := (t_1, t_2, \ldots, t_m)^T$. We suggest as one of various approaches of this study: (i) First to find the factor levels $\mathbf{x}_j^t := ((x_j^t)_1, (x_j^t)_2, \ldots, (x_j^t)_n)^T$ $(j = 1, 2, \ldots, m)$ corresponding to the ideal solutions t_j, i.e., $t_j := Y_j(\mathbf{x}_j^t)$ for each individual function f, (ii) then to compute the convex hull of these optimal solutions $\mathbf{x}_j^t$ $(j = 1, 2, \ldots, m)$ and determine some compromised solution $\bar{\mathbf{x}} := (\bar{x}_1, \bar{x}_2, \ldots, \bar{x}_n)^T$ which may not be the global one for the overall problem given in the previous chapter but a close one.

In other words, we firstly solve a representation problem of searching for an $m \times n$ design matrix

$$\mathbf{X}^t := (\mathbf{x}_1^t, \mathbf{x}_2^t, \ldots, \mathbf{x}_m^t)^T \tag{31}$$

by finding the zero of the system of $\mathbf{Y}(\mathbf{X}^t) - \mathbf{t}$, where $\mathbf{Y} = (Y_1, Y_2, \ldots, Y_m)^T$. Then we take the convex hull to obtain a compromised factor level $\bar{\mathbf{x}} := (\bar{x}_1, \bar{x}_2, \ldots, \bar{x}_n)^T$, i.e., the solution of $\bar{\mathbf{y}} := \mathbf{Y}(\bar{\mathbf{x}})$, where $\bar{\mathbf{y}} := (\bar{y}_1, \bar{y}_2, \ldots, \bar{y}_m)^T$ is a compromised solution in $\mathbf{y}$-space. We call this approach a *two-stage method*, because it is similar to the other *bilevel* approaches (Dempe 2002): First, we consider the *optimization problem* only in $\mathbf{y}$ as the lower level problem stated in Sect. 3; then, by introducing $\mathbf{x}$ into our analysis, we pass to the upper level of this problem, which contains a *representation problem*.

Let us recall that per y_j, we are in a compact interval $\mathbb{I}_j = [l_j, u_j]$ $(j = 1, 2, \ldots, m)$, i.e., feasible sets of the lower level problem and the individual functions $f_j(y_j)$ are continuous and nonsmooth. By the following assumption, we may think that the space $\mathbb{X} \subset \mathbb{R}^n$ of the factor variable $\mathbf{x}$ is compact, in fact, of the Cartesian product form $\mathbb{X} = [\mathbf{l}_x, \mathbf{u}_x] = \underset{i=1}{\overset{n}{\times}} \mathbb{X}_i$ with $\mathbb{X}_i := [l_{x_i}, u_{x_i}]$ $(i = 1, 2, \ldots, n)$ being compact intervals. In this case, $\mathbb{X}$ is a *parallelepiped* and, hence, for each $j = 1, 2, \ldots, m$, the image $Y_j(\mathbb{X})$ is again an interval which can be defined as our interval $\mathbb{I}_j$, i.e., $Y_j(\mathbb{X}) := \mathbb{I}$. We introduce

$$\mathbb{X}^{appr} := \mathrm{co}\{\mathbf{x}_1^t, \mathbf{x}_2^t, \ldots, \mathbf{x}_m^t\}, \tag{32}$$

where the points $\mathbf{x}_j^t$ $(j = 1, 2, \ldots, m)$ are solutions of the zero problems $Y_j(\mathbf{x}) - t_j = 0$ $(j = 1, 2, \ldots, m)$ together with the vector-valued condition $\mathbf{x} \in \mathbb{X}$, which can be represented further by 2^n linear inequality constraints. Altogether, we arrive at $2^n + m$ scalar-valued constraints. We note that $\mathbb{X}^{appr}$ is convex, in fact, a *polytope* and, hence, because of the convexity of $\mathbb{X}$, it holds $\mathbb{X}^{appr} \subseteq \mathbb{X}$. Here, we have a discrete structure in the entire $\mathbf{x}$-space, given by the vertices of $\mathbb{X}^{appr}$, and could further optimize (select) over the full polytope $\mathbb{X}^{appr}$ in that space. The weights may, e.g., come from the exponents given in the conventional DF. Indeed, we could choose $\bar{\mathbf{x}} = \sum_{j=1}^{m} w_j \mathbf{x}_j^t$. The main advantage of this coupling is that we would get an optimizer in that polytope within the full dimensions of $\mathbb{R}^n$. However, since still the variables are treated in a separated way, this new approach is just an approximation to our original problem. This approximation can be simplifying very much, because of the *joint* dependence of all the $y_j = Y_j(\mathbf{x})$ $(j = 1, 2, \ldots, m)$ on $\mathbf{x}$ and because of the *nonlinearity* and *nonconvexity* of the function f_j and Y_j $(j = 1, 2, \ldots, m)$. It can be motivated by *game theory* and introduces $\mathbb{X}^{appr}$ as a set of *compromise solutions*.

Another new opportunity is to further look for conditions to apply versions of the *Intermediate Value Theorem* directly in the *full* dimensions of the vector variable $\mathbf{y}$, rather than in each dimension with the difficulty of selecting the suitable optimizer in the $\mathbf{x}$-space then. In more general terms, we may also speak of the *Implicit Function Theorem*. Here, the structure of the functions Y_j, e.g., the relations between the x_i $(i = 1, 2, \ldots, n)$ and the y_j $(j = 1, 2, \ldots, m)$, is an important issue. We must have an (arcwise) connected domain of the vector-valued function $\mathbf{Y}(\mathbf{x})$ $(= (Y_1(\mathbf{x}), Y_2(\mathbf{x}), \ldots, Y_m(\mathbf{x}))^T)$, which we equate with $(\bar{y}_1, \bar{y}_2, \ldots, \bar{y}_m)^T$, and hence, of each of its components $Y_j(\mathbf{x})$ $(j = 1, 2, \ldots, m)$.

Let us summarize that this initial, pioneering and approximative approach has consisted of a separate consideration of the components y_j, combined with an enumeration (minimizing in a set of finitely many indices) along the pieces in each of these components, of a possible application of the Intermediate Value Theorem on the corresponding $Y_j(\mathbf{x})$ and, finally, of a polytope and selection argument in $\mathbb{X}^{appr}$, in order to find a compromise solution $\bar{\mathbf{x}}$ of the given problem.

In any case, what can be done is:

i. define a weighed sum of the components $Y_j(\mathbf{x})$, e.g., by the exponents that we can take from the DF as the weights, and to apply the Intermediate Value Theorem on corresponding real-valued function, then our zero problem of finding $\mathbf{x} = \bar{\mathbf{x}}$ looks, e.g., as follows:

$$(\sum_{j=1}^{m} w_j Y_j)(\mathbf{x}) = \sum_{j=1}^{m} w_j \bar{y}_j. \tag{33}$$

ii. approach the system of $2^n + m$ equations: $Y_j(\mathbf{x}) - \bar{y}_j = 0$ $(j = 1, 2, \ldots, m)$ and $\mathbf{x} \in \mathbb{X}$, and treat it with the help of the theory of Inverse Problems, e.g., by the *Inverse Function Theorem* or the *Implicit Function Theorem*. We select

$$\bar{\mathbf{x}} = \sum_{j=1}^{m} \hat{w}_j \mathbf{x}_j^t \quad (\text{where } \hat{w}_j \geq 0 \ (j = 1, 2, \ldots, m) \text{ and } \sum_{j=1}^{m} \hat{w}_j = 1), \tag{34}$$

e.g.,

$$\bar{\mathbf{x}} = \sum_{j=1}^{m} w_j \mathbf{x}_j^t, \ (\text{where } \hat{w}_j = w_j \ (j = 1, 2, \ldots, m)) \quad \text{or, especially}$$

$$\bar{\mathbf{x}} = \frac{1}{m} \sum_{j=1}^{m} \mathbf{x}_j^t, \ (\text{where } \hat{w}_j = 1 \text{ for all } (j = 1, 2, \ldots, m)). \tag{35}$$

5   Concluding Remarks

Many optimizaion approaches presented for DFs can be utilized in practice to find the root desirable solutions to multi-response optimization problems. Other approaches present opportunities for researches to improve performances. In this bookchapter, we elaborate the nondifferentiability issues of DFs of Derringer and Suich with a structural approach. A new multi-objective optimization method is suggested which is motivated by more than one nonddiferentiable point case of DFs. The We present the improvements in the MRO literature related with the response dependencies and robustness issues.

References

Akteke-Öztürk, B., Köksal, G., Weber, G.-W.: Nonconvex optimization of desirability functions using nonsmooth and global optimization approaches. Technical report, 01, Department of Industrial Engineering, Middle East Technical University

BARON, v. 8.1.5 (2010). www.gams.com/solver

Burachik, R.S., Gasimov, R.N., Ismayilova, N.A., Kaya, C.Y.: On a modified subgradient algorithm for dual problems via sharp augmented lagrangian. J. Global Optim. **34**(1), 55–78 (2006)

Chen, H.-W., Wong, W.K., Hongquan, X.: An augmented approach to the desirability function. J. Appl. Stat. **39**(3), 599–613 (2012)

Ch'ng, C.K., Quah, S.H., Low, H.C.: A new approach for multiple-response optimization. Qual. Eng. **17**, 621–626 (2005)

Clarke, F.: Optimization and Nonsmooth Analysis. SIAM's Classics in Applied Mathematics Series. SIAM, Philadelphia (1983)

Conn, A.R., Scheinberg, K., Vicente, L.N.: Introduction to Derivative-Free Optimization. MPS-SIAM Book Series on Optimization. SIAM, Philadelphia (2009)

Costa, N.R., Loureno, J.: Desirability function approach: a review and performance evaluation in adverse conditions. Chemometr. Intell. Lab Syst. **107**, 234–244 (2011)

CONOPT, v. 3.14S (2010). www.gams.com/solver

Del Castillo, E., Montgomery, D.C., McCarville, D.R.: Modified desirability functions for multiple response optimization. J. Qual. Technol. **28**(3), 337–345 (1996)

Dempe, S.: Foundations of Bilevel Programming. Kluwer Academic Publishers, Dordrecht (2002)

Demyanov, V.F., Rubinov, A.M.: Quasidifferentiable Calculus. Optimization Software. Publications Division, New York (1986)

Derringer, G., Suich, R.: Simultaneous optimization of several response variables. J. Qual. Technol. **12**, 214–219 (1980)

Derringer, G.: A balancing act, optimizing a products properties. Qual. Prog. **27**, 51–57 (1994)

DICOPT, v. 2x-C (2010) www.gams.com/solver

Drud, A.S.: CONOPT: a system for large scale nonlinear optimization, tutorial for CONOPT Subroutine Library, p. 16. ARKI Consulting and Development A/S, Bagsvaerd (1995b)

Dutta, J.: Generalized derivatives and nonsmooth optimization, a finite dimensional tour. TOP **13**(2), 185–314 (2005)

Ehrgott, M.: Multicriteria Optimization. Springer, Heidelberg (2005)

Fogliatto, F.S.: A survey of techniques for optimizing multiresponse experiments. In: Anais do XVIII ENEGEP Niteroi (1998)

Fuller D., Scherer, W.: The desirability function: Underlying assumptions and application implications. In: IEEE International Conference on Paper Presented at the Systems, Man, and Cybernetics, San Diego, CA (1998)

GAMS, v. 23.0.2 (2010). www.gams.com

Gasimov, R.N., Ustun, O.: Solving the quadratic assignment problem using F-MSG algorithm. J. Ind. Manag. Optim. **3**(2), 173–191 (2007)

Harrington Jr., E.C.: The desirability function. Ind. Qual. Control **21**, 494–498 (1965)

Jeong, I.J., Kim, K.J.: An interactive desirability function method to multiresponse optimization. Eur. J. Oper. Res. **195**(2), 412–426 (2008)

Khuri, A.I.: Multiresponse surface methodology. In: Ghosh, A., Rao, C.R. (eds.) Handbook of statistics: design and analysis of experiments, pp. 377–406. Elsevier, Amsterdam (1996)

Kim, K.J., Lin, D.: Simultaneous optimization of multiple responses by maximizing exponential desirability functions. Appl. Stat. **49**(C), 311–325 (2000)

Lasdon, L.S., Waren, A.D., Jain, A., Ratner, M.: Design and testing of a generalized reduced gradient code for nonlinear programming. ACM Trans. Math. Softw. **4**(1), 34–50 (1978)

Lemarechal, C.: Bundle-methods in nonsmooth optimization. In: Lemarechal, C., Mifflin, R. (eds.) Nonsmooth Optimization. Pergamon Press, Oxford (1978)

Logothetis, N., Wynn, H.P.: Quality Through Design. Oxford Science Publications, Clarendon Press, Oxford (1989)

Lundell, A., Westerlund, T.: Global optimization of mixed-integer signomial programming problems. In: Lee, J., Leyffer, S. (eds.) Mixed Integer Nonlinear Programming. The IMA Volumes in Mathematics and its Applications, pp. 349–369. Springer, New York (2012)

Miettinen, K.: Nonlinear Multiobjective Optimization. Kluwer Academic Publishers, Boston (1999)

Montgomery, D.C.: Design and Analysis of Experiments, 5th edn. Wiley, New York (2000)

Murphy, T., Tsui, K.-L., Allen, J.K.: A review of robust design methods for multiple responses. Res. Eng. Design **16**, 118–132 (2005)

Ozdaglar, A., Tseng, P.: Existence of global minima for constrained optimization. J. Optim. Theory Appl. **128**, 523–546 (2006)

Pardalos, P.M., Romeijn, H.E. (eds.): Handbook of Global Optimization, vol. 2. Kluwer Academic, Dordrecht (2002)

Park, K.S., Kim, K.J.: Optimizing multi-response surface problems: how to use multi-objective optimization techniques. IIE Trans. **37**(6), 523–532 (2005)

Ryoo, H.S., Sahinidis, N.V.: A branch-and-reduce approach to global optimization. J. Global Optim. **8**(2), 107–138 (1996)

Tawarmalani, M., Sahinidis, N.V.: Convexification and Global Optimization in Continuous and Mixed-Integer Nonlinear Programming: Theory, Algorithms, Software, and Applications. Nonconvex Optimization and Its Applications Series. Kluwer Academic Publishers, Boston (2002)

Wu, C.F.J., Hamada, M.: Experiments: Planning, Analysis, and Parameter Design Optimization. Wiley- Interscience, New York (2000)

Wu, F.-C.: Optimization of correlated multiple quality characteristics using desirability function. Qual. Eng. **17**(1), 119–126 (2004)

Wu, F.-C.: Robust design of nonlinear multiple dynamic quality characteristics. Comput. Ind. Eng. **56**, 1328–1332 (2009)

Kovach, J., Cho, B.R.: Development of a multidisciplinarymultiresponse robust designn optimization model. Eng. Optim. **40**(9), 805–819 (2008)

Modeling and Statistical Techniques
for Data Analysis

Analysis of Unreliable Single Server Queueing System with Hot Back-Up Server

Alexander Dudin[1]($\boxtimes$), Valentina Klimenok[1], and Vladimir Vishnevsky[2]

[1] Department of Applied Mathematics and Computer Science,
Belarusian State University, 220030 Minsk, Belarus
{dudin,klimenok}@bsu.by
[2] Institute of Control Sciences of Russian Academy of Sciences and Closed
Corporation "Information and Networking Technologies", Moscow, Russia
vishn@inbox.ru

Abstract. In this paper, we analyze an unreliable queueing system consisting of an infinite buffer and two heterogeneous servers. The main server (server 1) is unreliable, while the server 2 is considered as the reserve server and is assumed to be absolutely reliable. The service times have the PH-type (Phase-type) distribution. If both servers are able to provide the service, they serve a customer independently of each other. The service of a customer is completed when his/her service by any of two servers is finished. After the service completion, both servers immediately start the service of the next customer, if he/she presents in the system. If the system is idle, the servers wait for arrival of the new customer. The input flow is described by the $BMAP$ (Batch Markovian Arrival Process). Breakdowns arrive to the server 1 according to a MAP (Markovian Arrival Process). After breakdown occurrence, repair of the server starts. The repair time also has the PH-type (Phase-type) distribution. The customers, which meet the servers busy upon arrival, join a buffer. They will be picked up for the service according to the First-In-First-Out discipline. The customers arrived at the same batch are picked up for the service in random order. If a customer arriving from outside or from a buffer sees only server 2 ready for service while the server 1 is under repair, only server 2 starts the service of this customer. But if server 1 is repaired before service completion of this customer, server 1 immediately begins the service of this customer. For this model, we derive ergodicity condition, calculate the key performance measures of the system and derive an expression for the Laplace-Stieltjes transform of the sojourn time distribution of an arbitrary customer.

Keywords: Unreliable queueing system · Batch Markovian Arrival Process · Phase-type distribution · Stationary state distribution · Sojourn time distribution

1 Introduction

Queueing theory is an important branch of operations research. It provides a powerful mathematical tool for solving various problems of statical and

© Springer International Publishing Switzerland 2015
A. Plakhov et al. (Eds.): EmC-ONS 2014, CCIS 499, pp. 149–161, 2015.
DOI: 10.1007/978-3-319-20352-2_10

dynamical optimization of operation of real-life systems and processes. The most impressive applications of queueing theory are in the field of telecommunications. Early papers by A.K. Erlang at the beginning of the 20th century were inspired by the needs of *statical* control of telephone networks as the problems of minimization of equipment and staff of telephone companies subject to the fixed restrictions imposed on the indicators of quality of customers service (probability of dropping a call, waiting time, etc.). As an example of application of queueing theory to solution of the problem of *dynamic* control by the system operation (control by the robots (crawlers) that traverse the web and bring web pages to the indexing engine that updates the data base of a web search engine), the paper [1] can be mentioned.

During the last few years, intensive studies towards to improving the performance of wireless communication have been conducted within the development of the next generation (5G) networks. As it is mentioned in [2], one of the main directions of creating the ultra-high speed (up to 10 Gbit/s) and reliable wireless means of communication is the development of hybrid communication systems based on laser and radio-wave technologies. Because of the high practical need for hybrid communication systems, a considerable amount of studies of this class of systems have appeared recently. Some results of these studies are presented in [3–5]. Papers from [3] are focused mainly on the study of stationary reliability characteristics, methods and algorithms for optimal channel switching in hybrid systems by means of simulation.

Papers [4] and [5] are devoted to formulation and study of mathematical models of communication hybrid systems, consisting of a laser channel and a redundant IEEE 802.11n channel ("cold" or "hot" standby). Paper [4] deals with the so-called "cold" redundancy, in which the radio-wave link is assumed absolutely reliable (its work does not depend on the weather conditions) and it backs up the atmospheric optical (laser) communication link only in cases when the latter interrupts its functioning because of the unfavorable weather conditions. Upon the occurrence of favorable weather conditions the data packets begin to be transmitted over the FSO (Free Space Optical) channel. In the paper a statistical analysis of meteorological data for duration of the periods of favorable and unfavorable weather conditions is also carried out. In paper [5], the model of a hybrid communication channel with "hot" redundancy is considered, where the backup IEEE 802.11n channel is not idle and continuously transmits data along with the FSO channel, but, unlike the latter, at low speeds. The mathematical model of such hybrid channel is represented by a two-channel queuing system with a single unreliable server. The paper [2] presents results a further development of this study when a millimeter-wave (71–76 GHz, 81–86 GHz) radio channel is used as a backup one.

In the present paper, we consider queueing system suitable to model a hybrid communication channel with "hot" reserve under more general, in comparison with [5], assumptions about the pattern of arrival processes of customers and breakdowns and distribution of service times.

The rest of the paper consists of the following. In the next section, the mathematical model is described. The process of the system states as multi-dimensional continuous time Markov chain is described in Sect. 3. The generator of this Markov chain is presented. Section 4 is devoted to derivation of the ergodicity condition for this Markov chain and the brief description of the algorithm for computation of its stationary distribution. The vector probability generating function of this distribution and formulas for computation of some performance measures are obtained in Sect. 5. The sojourn time distribution in terms of Laplace-Stieltjes transform is obtained in Sect. 6. Section 7 concludes the paper.

2   Mathematical Model

We consider a queueing system consisting of an infinite buffer and two servers: the main working server (server 1) and the back-up (reserve) server (server 2). By default, we assume that the server 1 is high-speed but unreliable while the server 2 is low-speed but absolutely reliable. However, it should be mentioned that the presented analysis is implemented without any use of the fact that service rate at server 1 is higher than at the service rate at server 2.

Customers arrive into the system in accordance with the Batch Markovian Arrival Process ($BMAP$). The $BMAP$ is defined by the underlying process ν_t, $t \geq 0$, which is an irreducible continuous-time Markov chain with the finite state space $\{0, \ldots, W\}$, and the matrix generating function

$$\mathbf{D}(z) = \sum_{k=0}^{\infty} \mathbf{D}_k z^k, \ |z| \leq 1.$$

The batches of customers enter the system only at the epochs of the chain ν_t, $t \geq 0$, transitions. The $(W + 1) \times (W + 1)$ matrices $\mathbf{D}_k$, $k \geq 1$, (non-diagonal entries of the matrix $\mathbf{D}_0$) define the intensities of the process ν_t, $t \geq 0$, transitions which are accompanied by generating the k-size batch of customers. The matrix $\mathbf{D}(1)$ is an infinitesimal generator of the process ν_t, $t \geq 0$. The intensity (fundamental rate) of the $BMAP$ is defined as

$$\lambda = \boldsymbol{\theta} \mathbf{D}'(1)\mathbf{e}$$

where $\boldsymbol{\theta}$ is the unique solution of the system

$$\boldsymbol{\theta} \mathbf{D}(1) = \mathbf{0}, \ \boldsymbol{\theta}\mathbf{e} = 1,$$

and the intensity of batch arrivals is defined as $\lambda_b = \boldsymbol{\theta}(-\mathbf{D}_0)\mathbf{e}$. Here and in the sequel $\mathbf{e}(\mathbf{0})$ is a column (row) vector of appropriate size consisting of 1's (0's). The coefficient of variation, c_{var}, of intervals between batch arrivals is given by

$$c_{var}^2 = 2\lambda_b \boldsymbol{\theta}(-\mathbf{D}_0)^{-1}\mathbf{e} - 1$$

while the coefficient of correlation, c_{cor}, of intervals between successive batch arrivals is calculated as

$$c_{cor} = (\lambda_b \boldsymbol{\theta}(-\mathbf{D}_0)^{-1}(\mathbf{D}(1) - \mathbf{D}_0)(-\mathbf{D}_0)^{-1}\mathbf{e} - 1)/c_{var}^2.$$

For more information about the $BMAP$, its history, properties, special cases and related research see [6] and the survey paper by S. Chakravarthy [7].

If the servers are busy at an arrival epoch or the server 2 is busy while the server 1 is under repair, the customer is placed at the end of the queue in the buffer and is picked-up for a service later on, according the FIFO discipline. If an arriving customer or the first customer from the queue sees two servers idle and ready for service, he/she begins service at both servers. If server 1 is under repair and the server 2 is idle, the latter server begins the service of the customer. If the service of a customer is not finished until the end of repair period, server 1 immediately starts the service of the customer. The service of a customer is considered be completed when his/her service by any of two servers is finished.

The flow of breakdowns is defined as the MAP with an underlying process $\zeta_t, t \geq 0$, which takes values in the set $\{0, 1, \ldots, Z\}$ and is defined by the matrices $\mathbf{F}_0$ $\mathbf{F}_1$. The fundamental rate of this MAP is

$$\varphi = \phi \mathbf{F}_1 \mathbf{e}$$

where the row vector ϕ is the unique solution of the system

$$\phi(\mathbf{F}_0 + \mathbf{F}_1) = \mathbf{0}, \ \phi \mathbf{e} = 1.$$

When the server fails, the repair period starts immediately. Duration of this period has PH type distribution with an irreducible representation $(\tau, \mathbf{T})$. It means the following. Repair time is interpreted as the time until the continuous time Markov chain $\eta_t, t \geq 0$, with the state space $\{1, \ldots, R+1\}$ reaches the single absorbing state $R + 1$. Transitions of the chain $\eta_t, t \geq 0$, within the state space $\{1, \ldots, R\}$ are defined by the sub-generator $\mathbf{T}$ while the intensities of transitions into the absorbing state are defined by the vector $\mathbf{T}_0 = -\mathbf{T}\mathbf{e}$. At the service beginning epoch, the state of the process $m_t, t \geq 0$, is chosen within the state space $\{1, \ldots, R\}$ according to the probabilistic row vector τ. It is assumed that the matrix $\mathbf{T} + \mathbf{T}_0 \tau$ is an irreducible one. The repair rate is calculated as

$$æ = -(\tau \mathbf{T}^{-1} \mathbf{e})^{-1}.$$

For more information about the PH type distribution, its properties, partial cases, and suitability for approximation of a variety of probability distributions arising in modelling real-life systems see, e.g., [8].

Breakdowns arriving during the repair time are ignored by the system.

The service time of a customer by the kth server, $k = 1, 2$, has PH type distribution with an irreducible representation $(\beta^{(k)}, \mathbf{S}^{(k)})$. The service process on the kth server is directed by the Markov chain $m_t^{(k)}, t \geq 0$, with the state space $\{1, \ldots, M^{(k)}, M^{(k)}+1\}$ where $M^{(k)}+1$ is an absorbing state. The intensities of transitions into the absorbing state are defined by the vector $\mathbf{S}_0^{(k)} = -\mathbf{S}^{(k)}\mathbf{e}$. The service rates are calculated as

$$\mu^{(k)} = -[\beta^{(k)}(\mathbf{S}^{(k)})^{-1}\mathbf{e}]^{-1}, \ k = 1, 2.$$

3   Process of the System States

Let

- i_t be the number of customers in the system at the moment t, $i_t \geq 0$;
- $r_t = 0$, if server 1 is under repair at the moment t, $r_t = 1$, if server 1 is not broken, i.e., both the servers are fault-free at the moment t;
- $m_t^{(k)}$ be the state of the directing process of the service at the k-th busy server, $m_t^{(k)} = \overline{1, M^{(k)}}$, $k = 1, 2$;
- η_t be the state of the directing process of the repair time at the server 1, $\eta_t = \overline{1, R}$, if $r_t = 0$;
- ν_t and ζ_t be the states of the directing process of the $BMAP$ of customers and the MAP of breakdowns, correspondingly, $\nu_t = \overline{0, W}$, $\zeta_t = \overline{0, Z}$, at the epoch $t, t \geq 0$.

The process of the system states is described by the regular irreducible continuous time Markov chain, $\xi_t, t \geq 0$, with the state space

$$X = \{(0, 0, \eta, \zeta, \nu)\} \bigcup \{(0, 1, \zeta, \nu)\} \bigcup \{(i, 0, \eta, m^{(2)}, \zeta, \nu)\} \bigcup$$

$$\{(i, 1, m^{(1)}, m^{(2)}, \zeta, \nu)\}, \ i \geq 1, \ \eta = \overline{1, R}, \ m^{(k)} = \overline{1, M^{(k)}}, k = 1, 2,$$

$$\zeta = \overline{0, Z}, \ \nu = \overline{0, W}.$$

In the following, we will assume that the states of the chain $\xi_t, t \geq 0$, are ordered as follows. Under fixed values of (i, r), the states of the chain are enumerated in the lexicographic order. Denote the obtained ranked sets as $\Omega_{i,r}$, and arrange the state space X as follows:

$$\Omega_{0,0}, \ \Omega_{0,1}, \ \Omega_{1,0}, \ \Omega_{1,1}, \ \Omega_{2,0}, \ \Omega_{2,1}, \ \Omega_{3,0}, \ \Omega_{3,1}, \ldots.$$

Let $\mathbf{Q}_{ij}$, $i, j \geq 0$, be the matrices formed by intensities of the chain transition from the state corresponding to the value i of the component i_n to the state corresponding to the value j of this component. The following statement is true.

Lemma 1. *Infinitesimal generator Q of the Markov chain $\xi_t, t \geq 0$, has the block structure*

$$Q = \begin{pmatrix} \mathbf{Q}_{0,0} & \mathbf{Q}_{0,1} & \mathbf{Q}_{0,2} & \mathbf{Q}_{0,3} & \cdots \\ \mathbf{Q}_{1,0} & \mathbf{Q}_1 & \mathbf{Q}_2 & \mathbf{Q}_3 & \cdots \\ \mathbf{O} & \mathbf{Q}_0 & \mathbf{Q}_1 & \mathbf{Q}_2 & \cdots \\ \mathbf{O} & \mathbf{O} & \mathbf{Q}_0 & \mathbf{Q}_1 & \cdots \\ \vdots & \vdots & \vdots & \vdots & \ddots \end{pmatrix}$$

where the non-zero blocks of generator are of the following form:

$$\mathbf{Q}_{0,0} = \begin{pmatrix} \mathbf{T} \oplus (\mathbf{F}_0 + \mathbf{F}_1) \oplus \mathbf{D}_0 & \mathbf{T}_0 \otimes \mathbf{I}_{\bar{Z}} \otimes \mathbf{I}_{\bar{W}} \\ \boldsymbol{\tau} \otimes \mathbf{F}_1 \otimes \mathbf{I}_{\bar{W}} & \mathbf{F}_0 \oplus \mathbf{D}_0 \end{pmatrix},$$

$$Q_{0,k} = \begin{pmatrix} I_R \otimes \beta^{(2)} \otimes I_{\bar{Z}} \otimes D_k & O \\ O & \beta^{(1)} \otimes \beta^{(2)} \otimes I_{\bar{Z}} \otimes D_k \end{pmatrix}, \; k \geq 1,$$

$$Q_{1,0} = \begin{pmatrix} I_R \otimes S_0^{(2)} \otimes I_{\bar{Z}} \otimes I_{\bar{W}} & O \\ O & (S_0^{(1)} \otimes e_{M^{(2)}} + e_{M^{(1)}} \otimes S_0^{(2)}) \otimes I_{\bar{Z}} \otimes I_{\bar{W}} \end{pmatrix},$$

$$Q_0 = \begin{pmatrix} I_R \otimes S_0^{(2)} \beta^{(2)} \otimes I_{\bar{Z}} \otimes I_{\bar{W}} & O \\ O & \tilde{S} \otimes I_{\bar{Z}} \otimes I_{\bar{W}} \end{pmatrix},$$

$$Q_1 = \begin{pmatrix} T \oplus S^{(2)} \oplus (F_0 + F_1) \oplus D_0 & T_0 \beta^{(1)} \otimes I_{M^{(2)}} \otimes I_{\bar{Z}} \otimes I_{\bar{W}} \\ e_{M^{(1)}} T \otimes I_{M^{(2)}} \otimes F_1 \otimes I_{\bar{W}} & S^{(1)} \oplus S^{(2)} \oplus F_0 \oplus D_0 \end{pmatrix},$$

$$Q_k = \begin{pmatrix} I_R \otimes I_{M^{(2)}} \otimes I_{\bar{Z}} \otimes D_{k-1} & O \\ O & I_{M^{(1)}} \otimes I_{M^{(2)}} \otimes I_{\bar{Z}} \otimes D_{k-1} \end{pmatrix}, \; k \geq 2,$$

where

$$\tilde{S} = S_0^{(1)} \beta^{(1)} \otimes e_{M^{(2)}} \beta^{(2)} + e_{M^{(1)}} \beta^{(1)} \otimes S_0^{(2)} \beta^{(2)},$$

$\bar{W} = W + 1$, $\bar{Z} = Z + 1$, $\otimes$ *and* $\oplus$ *are the symbols of Kronecker product and sum of matrices, see, e.g., [9].*

The proof of the lemma is implemented by means of calculation of probabilities of transitions of the components of the Markov chain during a time interval having infinitesimal length.

Corollary 1. *The Markov chain* $\xi_t, t \geq 0$, *belongs to the class of continuous time quasi-Toeplitz Markov chains, see [10].*

Denote $\tilde{Q}(z) = \sum\limits_{k=1}^{\infty} Q_{0,k} z^k$, $Q(z) = \sum\limits_{k=0}^{\infty} Q_k z^k$, $|z| \leq 1$.

Corollary 2. *The matrix generating functions* $\tilde{Q}(z)$, $Q(z)$ *are of the form*

$$\tilde{Q}(z) = z \begin{pmatrix} I_R \otimes \beta^{(2)} \otimes I_{\bar{Z}} \otimes I_{\bar{W}} & O \\ O & \beta^{(1)} \otimes \beta^{(2)} \otimes I_{\bar{Z}} \otimes I_{\bar{W}} \end{pmatrix} +$$

$$\text{diag}\{I_{R\bar{Z}} \otimes (D(z) - D_0), \; I_{\bar{Z}} \otimes (D(z) - D_0)\}, \tag{1}$$

$$Q(z) = Q_0 + z \begin{pmatrix} T \oplus S^{(2)} \oplus (F_0 + F_1) \otimes I_{\bar{W}} & T_0 \beta^{(1)} \otimes I_{M^{(2)}} \otimes I_{\bar{Z}} \otimes I_{\bar{W}} \\ e_{M^{(1)}} T \otimes I_{M^{(2)}} \otimes F_1 \otimes I_{\bar{W}} & S^{(1)} \oplus S^{(2)} \oplus F_0 \otimes I_{\bar{W}} \end{pmatrix}$$

$$+ z \text{diag}\{I_{RM^{(2)}\bar{Z}} \otimes D(z), \; I_{M^{(1)}M^{(2)}\bar{Z}} \otimes D(z)\}. \tag{2}$$

Here $\text{diag}\{\bullet, \bullet\}$ *denotes the diagonal matrix with the diagonal blocks listed in the brackets.*

4   Ergodicity Condition. Stationary Distribution of the System States

Theorem 1. *The necessary and sufficient condition for ergodicity of the Markov chain ξ_t, $t \geq 0$, is the fulfillment of the inequality*

$$\lambda < \mathbf{x} \operatorname{diag}\{\mathbf{e}_R \otimes \mathbf{S}_0^{(2)} \otimes \mathbf{e}_{\bar{Z}}, (\mathbf{S}_0^{(1)} \otimes \mathbf{e}_{M^{(2)}} + \mathbf{e}_{M^{(1)}} \otimes \mathbf{S}_0^{(2)}) \otimes \mathbf{e}_{\bar{Z}}\}\mathbf{e} \qquad (3)$$

where the vector $\mathbf{x}$ is the unique solution of the system of linear algebraic equations

$$\mathbf{x}\mathbf{A} = \mathbf{0}, \quad \mathbf{x}\mathbf{e} = 1 \qquad (4)$$

where

$$\mathbf{A} = \begin{pmatrix} \mathbf{T} \oplus \mathbf{S}^{(2)} \oplus (\mathbf{F}_0 + \mathbf{F}_1) + \mathbf{I}_R \otimes \mathbf{S}_0^{(2)} \boldsymbol{\beta}^{(2)} \otimes \mathbf{I}_{\bar{Z}} & \mathbf{T}_0 \boldsymbol{\beta}^{(1)} \otimes \mathbf{I}_{M^{(2)}} \otimes \mathbf{I}_{\bar{Z}} \\ \mathbf{e}_{M^{(1)}} \boldsymbol{\tau} \otimes \mathbf{I}_{M^{(2)}} \otimes \mathbf{F}_1 & \tilde{\mathbf{S}} \otimes \mathbf{I}_{\bar{Z}} + \mathbf{S}^{(1)} \oplus \mathbf{S}^{(2)} \oplus \mathbf{F}_0 \end{pmatrix}.$$

Proof. It follows from [10], that a necessary and sufficient condition for existence of the stationary distribution of the chain ξ_t, $t \geq 0$, can be formulated in terms of the matrix generating function $\mathbf{Q}(z)$ and has the form of the inequality

$$\mathbf{y}\mathbf{Q}'(1)\mathbf{e} < 1, \qquad (5)$$

where the row vector $\mathbf{y}$ is the unique solution of the system of linear algebraic equations

$$\mathbf{y}\mathbf{Q}(1) = \mathbf{0}, \quad \mathbf{y}\mathbf{e} = 1. \qquad (6)$$

Let the vector $\mathbf{y}$ be of the form

$$\mathbf{y} = \mathbf{x} \otimes \boldsymbol{\theta} \qquad (7)$$

where $\boldsymbol{\theta}$ is the vector of the stationary distribution of the $BMAP$ underlying process ν_t, $t \geq 0$, and $\mathbf{x}$ is a stochastic vector. Substituting expression (7) into (6), we verify that $\mathbf{y}$ is the unique solution of system (6) if the vector $\mathbf{x}$ satisfies system (4).

Substituting the vector $\mathbf{y}$ in the form (7) into (5) and using expression (2) to calculate the derivative $\mathbf{Q}'(1)$, we reduce inequality (5) to the form (3).

The theorem is proved.

Remark 1. Intuitive explanation of stability condition (3) is as follows. The left hand side of inequality (3) is the rate of customers arriving into the system. The right hand side of the inequality is a rate of customers leaving the system after service under overload condition. It is obvious that in steady state the former rate must be less that the latter one.

Corollary 3. *In the case of stationary Poisson flow of breakdowns and expo-nential distribution of service and repair times, ergodicity condition (3)-(4) is reduced to the following inequality:*

$$\lambda < \mu_2 + \frac{\text{æ}}{\text{æ} + \varphi}\mu_1. \tag{8}$$

Proof. In the case under consideration, the vector **x** consists of two components, say, $\mathbf{x} = (x_1, x_2)$. It is easy to see that inequality (3) is reduced to the following one:

$$\lambda < x_1\mu_2 + x_2(\mu_1 + \mu_2). \tag{9}$$

System (4) is written as

$$\begin{cases} -x_1\text{æ} + x_2\varphi = 0, \\ x_1\text{æ} - x_2\varphi = 0, \\ x_1 + x_2 = 1. \end{cases} \tag{10}$$

Relations (9)-(10) imply inequality (8). The corollary is proved.

In what follows we assume inequality (3) be fulfilled.

Denote the stationary state probabilities of the chain ξ_t, $t \geq 0$, by

$$p(0, 0, \eta, \zeta, \nu), \ p(0, 1, \zeta, \nu), \ p(i, 0, \eta, m^{(2)}, \zeta, \nu),$$

$$p(i, 1, m^{(1)}, m^{(2)}, \zeta, \nu), i \geq 1, \ \eta = \overline{1, R}, \ m^{(k)} = \overline{1, M^{(k)}}, k = 1, 2,$$

$$\zeta = \overline{0, Z}, \ \nu = \overline{0, W}.$$

Let us enumerate the steady state probabilities in accordance with the introduced above order of the states of the chain and form the row vectors $\mathbf{p}_i$ of steady state probabilities corresponding the value i of the first component of the Markov chain, $i \geq 0$.

To calculate the vectors $\mathbf{p}_i$, $i \geq 0$, we use the numerically stable algorithm, see [10], which has been elaborated for calculating the stationary distribution of multi-dimensional continuous time quasi-Toeplitz Markov chains. The derivation of this algorithm is based on censoring technique and the algorithm consists of the next principal steps.

Algorithm.

1. Calculate the matrix **G** as the minimal nonnegative solution of the non-linear matrix equation

$$\sum_{n=0}^{\infty} \mathbf{Q}_n\mathbf{G}^n = \mathbf{O}.$$

2. Calculate the matrix $\mathbf{G}_0$ from the equation

$$\mathbf{Q}_{1,0} + \sum_{n=1}^{\infty} \mathbf{Q}_{1,n}\mathbf{G}^{n-1}\mathbf{G}_0 = \mathbf{O}$$

whence it follows that

$$\mathbf{G}_0 = -(\sum_{n=1}^{\infty} \mathbf{Q}_{1,n}\mathbf{G}^{n-1})^{-1}\mathbf{Q}_{1,0}.$$

3. Calculate the matrices $\bar{\mathbf{Q}}_{i,l}$, $l \geq i, i \geq 0$, using the formulae

$$
\bar{\mathbf{Q}}_{i,l} = \begin{cases} \mathbf{Q}_{0,l} + \sum\limits_{n=l+1}^{\infty} \mathbf{Q}_{0,n}\mathbf{G}_{n-1}\mathbf{G}_{n-2}\ldots\mathbf{G}_l, & i=0, l \geq 0, \\ \mathbf{Q}_{l-i} + \sum\limits_{n=l+1}^{\infty} \mathbf{Q}_{n-i}\mathbf{G}_{n-1}\mathbf{G}_{n-2}\ldots\mathbf{G}_l, & i \geq 1, l \geq i, \end{cases}
$$

where $\mathbf{G}_i = \mathbf{G}$, $i \geq 1$.

4. Calculate the matrices $\mathbf{\Phi}_l$ using the recurrent formulae

$$
\mathbf{\Phi}_0 = \mathbf{I}, \mathbf{\Phi}_l = \sum_{i=0}^{l-1} \mathbf{\Phi}_i \bar{\mathbf{Q}}_{i,l}(-\bar{\mathbf{Q}}_{l,l})^{-1}, l \geq 1.
$$

5. Calculate the vector $\mathbf{p}_0$ as the unique solution of the system

$$
\mathbf{p}_0(-\bar{\mathbf{Q}}_{0,0}) = \mathbf{0}, \ \mathbf{p}_0 \sum_{l=0}^{\infty} \mathbf{\Phi}_l \mathbf{e} = 1.
$$

6. Calculate the vectors $\mathbf{p}_l$, $l \geq 1$, as follows:

$$
\mathbf{p}_l = \mathbf{p}_0 \mathbf{\Phi}_l, \ l \geq 1.
$$

The proposed algorithm is numerically stable because all matrices involved into the algorithm are non-negative.

5   Vector Generating Function of the Stationary Distribution. Performance Measures

Having the stationary distribution of the system states $\mathbf{p}_i$, $i \geq 0$, been calculated we can find a number of stationary performance measures of the considered system. When calculating the performance measures, the following result will be useful, especially in the case when the distribution $\mathbf{p}_i$, $i \geq 0$, is heavy tailed.

Lemma 2. *The vector generating function* $\mathbf{P}(z) = \sum\limits_{i=1}^{\infty} \mathbf{p}_i z^i$, $|z| \leq 1$, *satisfies the following equation:*

$$
\mathbf{P}(z)\mathbf{Q}(z) = z[\mathbf{p}_1\mathbf{Q}_0 - \mathbf{p}_0\tilde{\mathbf{Q}}(z)]. \tag{11}
$$

Remark 2. Equation (11) is indeed the functional-differential equation because it is equivalent to equation

$$
\mathbf{P}(z)\mathbf{Q}(z) = z\left[\frac{d\mathbf{P}(z)}{dz}|_{z=0}\mathbf{Q}_0 - \mathbf{p}_0\tilde{\mathbf{Q}}(z)\right].
$$

In particular, formula (11) can be used to calculate the value of the generating function $\mathbf{P}(z)$ and its derivatives at the point $z = 1$ without the calculation of infinite sums. Having these derivatives been calculated, we will able to find moments of the number of customers in the system and some others performance measures of the system. The problem of calculating the value of the vector generating function $\mathbf{P}(z)$ and its derivatives at the point $z = 1$ from Eq. (11) is non-trivial one because the matrix $\mathbf{Q}(z)$ is singular at the point $z = 1$.

Let us denote $f^{(n)}(z)$ the nth derivative of the function $f(z)$, $n \geq 1$, and $f^{(0)}(z) = f(z)$.

Corollary 4. *The mth, $m \geq 0$, derivatives of the vector generating function $\mathbf{P}(z) = \sum\limits_{i=1}^{\infty} \mathbf{p}_i z^i$, $|z| \leq 1$, at the point $z = 1$ (so called vector factorial moments) are recursively calculated as the solution of the system of linear algebraic equations*

$$
\begin{cases}
\mathbf{P}^{(m)}(1)\mathbf{Q}(1) = \mathbf{H}^{(m)}(1) - \sum\limits_{l=0}^{m-1} C_m^l \mathbf{P}^{(l)}(1)\mathbf{Q}^{(m-l)}(1), \\
\mathbf{P}^{(m)}(1)\mathbf{Q}'(1)\mathbf{e} = \frac{1}{m+1}[\mathbf{H}^{(m+1)}(1) - \sum\limits_{l=0}^{m-1} C_{m+1}^l \mathbf{P}^{(l)}(1)\mathbf{Q}^{(m+1-l)}(1)]\mathbf{e}.
\end{cases}
\tag{12}
$$

where

$$
\mathbf{H}^{(m)}(1) = \begin{cases}
\mathbf{p}_1\mathbf{Q}_0 - \mathbf{p}_0\tilde{\mathbf{Q}}(1), & m = 0, \\
\mathbf{p}_1\mathbf{Q}_0 - \mathbf{p}_0\tilde{\mathbf{Q}}(1) - \mathbf{p}_0\tilde{\mathbf{Q}}'(1), & m = 1, \\
-\mathbf{p}_0[m\tilde{\mathbf{Q}}^{(m-1)}(1) + \mathbf{Q}^{(m)}(1)], & m > 1,
\end{cases}
$$

and the derivatives $\mathbf{Q}^{(m)}(1)$, $\tilde{\mathbf{Q}}^{(m)}(1)$ are easily calculated using formulas (1)-(2).

The proof of the corollary is based on the technique very similar to the one outlined in paper [11] and is omitted here.

Further, we list some performance measures of the system under consideration.

- Throughput of the system (maximal number of the customers that can be processed during an unit of time) is defined by the right hand side of inequality (3).
- Mean number of customers in the system

$$
L = \mathbf{P}^{(1)}(1)\mathbf{e}.
$$

- Variance of the number of customers in the system

$$
V = \mathbf{P}^{(2)}(1)\mathbf{e} + L - L^2.
$$

- The share of time when the server 1 is fault-free

$$
P_1 = \mathbf{P}(1)\mathrm{diag}\{\mathbf{O}_{RM^{(2)}\bar{Z}\bar{W}}, \mathbf{I}_{M^{(1)}M^{(2)}\bar{Z}\bar{W}}\}\mathbf{e} + \mathbf{p}_0\mathrm{diag}\{\mathbf{O}_{R\bar{Z}\bar{W}}, \mathbf{I}_{\bar{Z}\bar{W}}\}\mathbf{e}.
$$

- The share of time when the server 1 is under repair

$$
P_0 = \mathbf{P}(1)\mathrm{diag}\{\mathbf{I}_{RM^{(2)}\bar{Z}\bar{W}}, \mathbf{O}_{M^{(1)}M^{(2)}\bar{Z}\bar{W}}\}\mathbf{e} + \mathbf{p}_0\mathrm{diag}\{\mathbf{I}_{R\bar{Z}\bar{W}}, \mathbf{O}_{\bar{Z}\bar{W}}\}\mathbf{e}.
$$

6   Sojourn Time Distribution

Let $V(x)$ be the stationary distribution function of the sojourn time of an arbitrary customer in the system. Denote the Laplace - Stieltjes transform of this function as $v(s) = \int_0^\infty e^{-sx} dV(x)$, $Re\ s \geq 0$.

Theorem 2. *The Laplace - Stieltjes transform of the sojourn time stationary distribution is calculated as*

$$v(s) = \lambda^{-1} \{ \mathbf{p}_0 \sum_{k=1}^\infty \mathbf{Q}_{0,k} \sum_{l=1}^k \boldsymbol{\Psi}^l(s) + \sum_{i=1}^\infty \mathbf{p}_i \sum_{k=2}^\infty \mathbf{Q}_k \sum_{l=1}^{k-1} \boldsymbol{\Psi}^{i+l}(s) \} \mathbf{e} \qquad (13)$$

where

$$\boldsymbol{\Psi}(s) = (sI - \hat{\mathbf{Q}})^{-1} \mathbf{Q}_0, \quad \hat{\mathbf{Q}} = \mathbf{Q}(1) - \mathbf{Q}_0. \qquad (14)$$

Proof. The proof is based on the probabilistic interpretation of the Laplace-Stieltjes transform. We assume that, independently on the system operation, the stationary Poisson input of so called catastrophes arrives. Let s, $s > 0$ be the intensity of this flow. Then, the Laplace-Stieltjes transform $v(s)$ can be interpreted as the probability of no catastrophe arrival during the sojourn time of an arbitrary customer. It allows to derive the expression for $v(s)$ by means of probabilistic reasonings.

Let us assume that at the moment of the beginning of a customer service the initial phases of service time at the servers are already determined. Then the matrix of probabilities of no catastrophes arrival during the service time of this customer (and corresponding transitions of the finite components of the Markov chain ξ_t, $t \geq 0$,) is evidently calculated as

$$\tilde{\boldsymbol{\Psi}}(s) = \int_0^\infty e^{(-sI + \hat{\mathbf{Q}})t} \mathbf{Q}_{1,0} dt = (sI - \hat{\mathbf{Q}})^{-1} \mathbf{Q}_{1,0}, \qquad (15)$$

if at the departure epoch there are no customers in the queue, and

$$\boldsymbol{\Psi}(s) = \int_0^\infty e^{(-sI + \hat{\mathbf{Q}})t} \mathbf{Q}_0 dt = (sI - \hat{\mathbf{Q}})^{-1} \mathbf{Q}_0, \qquad (16)$$

if at the departure epoch there are customers in the queue.

Note, that $\tilde{\boldsymbol{\Psi}}(s)\mathbf{e} = \boldsymbol{\Psi}(s)\mathbf{e}$ because $\mathbf{Q}_{1,0}\mathbf{e} = \mathbf{Q}_0\mathbf{e}$. So, the matrix $\tilde{\boldsymbol{\Psi}}(s)$ does not appear in the formula (13).

Assuming that an arbitrary customer arriving in a group of size k is placed on the jth position, $j = \overline{1, k}$, with probability $1/k$ and using the law of total probability, we obtain the following expression

$$v(s) = \mathbf{p}_0 \sum_{k=1}^\infty \frac{k}{\lambda} \mathbf{Q}_{0,k} \sum_{l=1}^k \frac{1}{k} \boldsymbol{\Psi}^l(s)\mathbf{e} + \sum_{i=1}^\infty \mathbf{p}_i \sum_{k=2}^\infty \frac{k-1}{\lambda} \mathbf{Q}_k \sum_{l=1}^{k-1} \frac{1}{k-1} \boldsymbol{\Psi}^{i+l}(s)\mathbf{e}. \qquad (17)$$

Formula (13) immediately follows from formula (17). The theorem is proved.

Remark 3. Expression (13) contains the infinite sums. But, it can be further simplified. E.g. expression

$$\sum_{k=1}^{\infty} \mathbf{Q}_{0,k} \sum_{l=1}^{k} \mathbf{\Psi}^l(s)$$

can be transformed as

$$\sum_{k=1}^{\infty} \mathbf{Q}_{0,k} \sum_{l=1}^{k} \mathbf{\Psi}^l(s) = (\tilde{Q}(1) - \tilde{Q}(\mathbf{\Psi}(s)))\mathbf{\Psi}(s)(I - \mathbf{\Psi}(s))^{-1}.$$

However, formula (13) is quite suitable for direct computations, especially if the $BMAP$ has a final support, i.e., only a finite number of matrices D_k are non-zero.

Corollary 5. *Mean sojourn time $\bar{v}$ of an arbitrary customer in the system is calculated as*

$$\bar{v} = -\lambda^{-1}[\mathbf{p}_0 \sum_{k=1}^{\infty} \mathbf{Q}_{0,k} \sum_{l=1}^{k} \sum_{m=0}^{l-1} \mathbf{\Psi}^m(0)+$$

$$\sum_{i=1}^{\infty} \mathbf{p}_i \sum_{k=2}^{\infty} \mathbf{Q}_k \sum_{l=1}^{k-1} \sum_{m=0}^{i+l-1} \mathbf{\Psi}^m(0)]\mathbf{\Psi}'(0)\mathbf{e} \qquad (18)$$

where

$$\mathbf{\Psi}'(0) = -[\hat{\mathbf{Q}}(1)]^{-2}\mathbf{Q}_0.$$

Proof. To obtain formula (18), we differentiate Eq. (16) at the point $s = 0$ and use the well-known relation $\bar{v} = -v'(0)$. Further, we reduce the obtained expression to the form (18) using the fact that the matrix $\mathbf{\Psi}(0)$ is a stochastic one.

7 Conclusion

We considered queueing model having two heterogeneous servers. One server is the main server, another one is reserved (back-up) server. The main, high speed, server is unreliable. The reserved, low speed, server is reliable. When the main server is available, it provides the service in parallel with the reserved server (this means that the reserved server is in a hot reserve). When the main server is broken, the service is performed only by the reserved server. Under quite general assumptions about arrival processes of customers and breakdowns and about distribution of service and repair times, we described behavior of the system by the multi-dimensional Markov chain that is a quasi-Toeplitz chain. Using known results for such a type of chains, we derived stability condition of this system, computed stationary distribution of the system states, presented expression for the vector generating function of this distribution and the recursive procedure for calculation of factorial moments of this distribution, as well as expressions for some performance measures of the system, obtained Laplace-Stieltjes distribution of sojourn time of an arbitrary customer and the average sojourn time.

Results can be used for development of software useful for performance evaluation and capacity planning of hybrid communication systems having high-speed laser channel as the main one and a radio-channel as the reserved one. Optimization problems related to the suitable choice of a bandwidth of a laser channel and a radio-channel, taking into account various service level objectives and energy saving issues, can be solved based on the presented results.

Acknowledgments. The research is supported by the Russian Foundation for Basic Research (grant No. 14-07-90015) and the Belarusian Republican Foundation for Fundamental Research (grant No. F14R-126).

References

1. Avrachenkov, K., Dudin, A., Klimenok, V., Nain, P., Semenova, O.: Optimal threshold control by the robots of web search engines with obsolescence of documents. Comput. Netw. **55**, 1880–1893 (2011)
2. Vishnevsky, V., Kozyrev, D., Semenova, O.V.: Redundant queueing system with unreliable servers. In: Proceedings of the 6th International Congress on Ultra Modern Telecommunications and Control Systems and Workshops (ICUMT), pp. 383–386. Moscow (2014)
3. Arnon, S., Barry, J., Karagiannidis, G., Schober, R., Uysal, M. (eds.): Advanced Optical Wireless Communication Systems. Cambridge University Press, Cambridge (2012)
4. Sharov, S.Y., Semenova, O.V.: Simulation model of wireless channel based on FSO and RF technologies. Distributed Computer and Communication Networks. Theory and Applications (DCCN-2010), pp. 368–374 (2010)
5. Vishnevsky, V.M., Semenova, O.V., Sharov, S.Y.: Modeling and analysis of a hybrid communication channel based on free-space optical and radio-frequency technologies. Autom. Remote Control **72**, 345–352 (2013)
6. Lucantoni, D.M.: New results on the single server queue with a batch Markovian arrival process. Commun. Stat.-Stoch. Models **7**, 1–46 (1991)
7. Chakravarthy, S.R.: The batch markovian arrival process: a review and future work. In: Krishnamoorthy, A., et al. (eds.) Advances in Probability Theory and Stochastic Processes, pp. 21–49. Notable Publications, NJ (2001)
8. Neuts, M.: Matrix-geometric Solutions in Stochastic Models - An Algorithmic Approach. Johns Hopkins University Press, USA (1981)
9. Graham, A.: Kronecker Products and Matrix Calculus with Applications. Ellis Horwood, Cichester (1981)
10. Klimenok, V.I., Dudin, A.N.: Multi-dimensional asymptotically quasi-Toeplitz Markov chains and their application in queueing theory. Queueing Systems **54**, 245–259 (2006)
11. Dudin, A., Klimenok, V.: Moon Ho Lee: recursive formulas for the moments of queue length in the $BMAP/G/1$ queue. IEEE Commun. Lett. **13**, 351–353 (2009)

Identification of DNA CpG Islands Using Inter-dinucleotide Distances

Vera Afreixo[1,2,4](✉), Carlos A.C. Bastos[2,3], João M.O.S. Rodrigues[2,3], and Raquel M. Silva[1,2]

[1] CIDMA - Center for Research and Development in Mathematics and Applications, University of Aveiro, 3810-193 Aveiro, Portugal
[2] IEETA - Institute of Electronics and Telematics Engineering of Aveiro, University of Aveiro, 3810-193 Aveiro, Portugal
vera@ua.pt
[3] Department of Electronics Telecommunications and Informatics, University of Aveiro, 3810-193 Aveiro, Portugal
[4] Institute for Research in Biomedicine - iBiMED, University of Aveiro, 3810-193 Aveiro, Portugal

Abstract. In this study we set to explore the potentialities of the inter-genomic symbols distance for finding CpG islands in DNA sequences. We explore the distance distributions of the inter CpG and SS distance in the independent nucleotide context (reference). We confront the empirical results from the complete human genome, CpG islands and non CpG islands, with the corresponding reference results.

We propose a model to discriminate CpG islands based on some statistical properties of the inter-dinucleotide distances distributions in DNA sequences. The results of this exploratory study suggest that inter-SS symbols distance has high ability to discriminate CpG islands.

1 Introduction

CpG islands (CGIs) are relatively small segments (ranging from a few hundred bases to several kilobases in length) of high CpG density where cytosines are generally unmethylated.

In this work we focus our analysis in CpG dinucleotide distance distribution vs all CG dinucleotide distance distributions.

For CpG islands detection there are several published algorithms/software. Many of the current software applications for CpG islands detection implement the definition of Gardiner-Garden and Frommer [4]. An extension was proposed by Takai and Jones [10] and several authors consider this algorithm one of the gold standard algorithms for CpG detection [8]. There are other well known algorithms for CpG islands detection. For example, CpGcluster identifies clusters of CpG dinucleotides with statistical significance, and this approach presents very good results [6,7]. The algorithms based on HMM (Hidden Markov Models) also

© Springer International Publishing Switzerland 2015
A. Plakhov et al. (Eds.): EmC-ONS 2014, CCIS 499, pp. 162–172, 2015.
DOI: 10.1007/978-3-319-20352-2_11

show good performance for CpG island detection [2] and are implemented, for example, in Geneious (Biomatters, available from www.geneious.com) sequence analysis software.

CpG dinucleotides are very frequent in CpG islands. Why is so important to detect CpG islands? CpG islands are usually associated with DNA regulation. In particular, many genes in mammalian genomes have CpG islands associated with the start of the gene (promoter regions) and transcription initiation.

What is so special about CpGs relative to the other 15 possible dinucleotides in DNA? CpGs are the sites where methylation takes place, and ∼80 % of them are methylated at position 5 of the cytosine ring in humans and mice.

However, CpGs may remain nonmethylated at CpG islands, despite their local abundance. For example, in mammals, methylating the cytosine of a CpG island can turn off expression of the associated gene (silenced), a mechanism that is part of a larger field of science studying gene regulation that is called epigenetics. In some human cancers, some "normal" genes are silenced in this way. In contrast, the hypomethylation of CpG sites has been associated with the over-expression of oncogenes within cancer cells [1,3].

In this paper, we focus on the CpG island properties in the context of inter-CpG/inter-SS distance distributions comparing the results with non CpG island and independent symbol model scenario. We apply stochastic process concepts to drive SS and CpG distance distributions in random sequence with independent symbols. The paper is divided into four sections. Section 2 presents the proposed methods: the inter-CpG/inter-SS distances and the corresponding reference distributions. Section 3 deals with the characterization of CpG islands, of non CpG islands and of the complete genome in *Homo sapiens* by inter-symbols distance distribution. Finally, Sect. 4 draws the conclusions of the paper.

2   Methods

2.1   Inter CpG Distance

We start by extracting the distance sequences. Consider, for example, the following sequence $CGATGCGACCGAATT\ldots$. To obtain the inter CpG distance: the nucleotide sequences are scanned for CpG dinucleotide $CG---CG--CG-$ $---\ldots$; the positions occupied by 'C' are recorded $x_1 = 1, x_2 = 6, x_3 = 10, \ldots$; and the CpG distance sequence is computed as the difference between those positions $d = 5, 4, \ldots$ $(d_i = x_{i+1} - x_i)$. From the CpG distance sequence we can obtain the empirical distance distribution.

To evaluate the exceptional behavior of CpG distances we derive the distance distribution under the assumption of independence between nucleotides. In particular, we will explore the probability distribution of the inter CpG distances under the independence of nucleotides. Note that, the dinucleotide counts are not independent even when we are under the independent nucleotide model (two adjacent dinucleotides share one nucleotide in common).

Under the nucleotide independence assumption the distance distribution can be found by analysing the state diagram in Fig. 1. Note that each distance is related with the number of steps to achieve success (CpG).

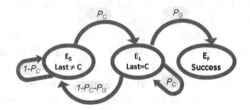

Fig. 1. State diagram for inter CpG distance.

Consider D the random variable which represents the inter CpG distance under the nucleotide independence assumption. Suppose that we are interested in characterizing the distance five (the first distance of our previous example). The probability of obtaining a distance equal to five is equal to the probability of arriving at the final state in 5 steps after starting in state 0. We can obtain the distance 5 using different paths. And the probability of distance 5 is the sum of the probabilities of all different paths. However, this approach results in a cumbersome description of the probability distribution. We explore a recursive form attending to the law of total probability, note that the probability of obtaining the distance 5 can be described by a sum: the probability of occurrence of a C (p_C) multiplied by the probability of obtaining the distance 4 plus the probability of not appearing a C multiplied by the probability of obtaining the distance 4.

In general, the probability of obtaining distance d is equal to the probability of occurrence of a C multiplied by the probability of obtain the distance $d-1$ plus the probability of not appearing a C times the probability of obtaining the distance $d-1$. The probability of obtain distance 1, i.e. obtaining the success (CpG) in one step, is zero.

$$P(D = d|E_0) = \begin{cases} 0 & , d = 1 \\ p_C\, P(D = d - 1|E_1) + (1 - p_C)\, P(D = d - 1|E_0) & , d \geq 2 \end{cases} \quad (1)$$

If we are in state E_1 we can define also a recursive expression to obtain the distance probability

$$P(D = d|E_1) = \begin{cases} p_G & , d = 1 \\ p_C\, P(D = d - 1|E_1) + (1 - p_C - p_G)\, P(D = d - 1|E_0) & , d \geq 2 \end{cases} \quad (2)$$

where p_G represents the probability of occurrence of a G.

The state diagram can be described by the following transition matrix of states E_0, E_1 and E_F

$$\mathbf{P} = \left[\begin{array}{cc|c} 1 - p_C & p_C & 0 \\ 1 - p_C - p_G & p_C & p_G \\ \hline 0 & 0 & 1 \end{array} \right] = \left[\begin{array}{c|c} \mathbf{Q} & \mathbf{R} \\ \hline \mathbf{0} & I \end{array} \right] \quad (3)$$

Recalling some concepts of Markov chains we can find the expected value of variable D and a closed form to the probability function. If we have an absorbing Markov chain we can obtain the corresponding fundamental matrix and the mean time to absorption [5]. $\mathbf{P}$ is an absorbing Markov chain for which the fundamental matrix is given by

$$\mathbf{F} = \mathbf{Q}^{-1} \begin{bmatrix} 1-(1-p_C) & -p_C \\ -(1-p_C-p_G) & 1-p_C \end{bmatrix}^{-1} \tag{4}$$

For computing the distance distribution we must consider E_0 as the first state. Thus, the time to absorption is the sum of values of $\mathbf{F}$ first row: $F_{11}+F_{12} = 1/(p_C\,p_G)$. This value is, naturally, the average of variable D.

Under an independent symbol model the probability distribution of the CpG distances can be written in matrix form as,

$$\begin{bmatrix} P(D=1|E_0) \\ P(D=1|E_1) \end{bmatrix} = \begin{bmatrix} 0 \\ p_G \end{bmatrix} \tag{5}$$

$$\begin{bmatrix} P(D=d|E_0) \\ P(D=d|E_1) \end{bmatrix} = \mathbf{Q} \begin{bmatrix} P(D=d-1|E_0) \\ P(D=d-1|E_1) \end{bmatrix}, \ d > 1. \tag{6}$$

And we can also present the closed form, concluding a kind of geometric distribution,

$$\begin{bmatrix} P(D=d|E_0) \\ P(D=d|E_1) \end{bmatrix} = \mathbf{Q}^{d-1} \begin{bmatrix} 0 \\ p_G \end{bmatrix}. \tag{7}$$

2.2   Inter SS Distance

In a procedure similar to that used to find the inter CpG distance distribution, we can also define the inter SS distance. Here, we follow the standard IUPAC (International Union of Pure and Applied Chemistry) nucleotide codes, where $S = \{C, G\}$ and $SS = \{CC, CG, GC, GG\}$. The nucleotide sequences are scanned for SS dinucleotide, the positions occupied by the first S are recorded and the SS sequence distance is computed, which we call SS sequence distance (Fig. 2 presents the corresponding state diagram). The random variable which represents the inter SS distance is denoted by D_S.

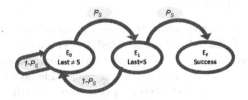

Fig. 2. State diagram for inter SS distance.

The state diagram can be described by the following transition matrix of states E_0, E_1 and E_F

$$\mathbf{P}_S = \begin{bmatrix} 1 - p_S & p_S & 0 \\ 1 - p_S & 0 & p_S \\ 0 & 0 & 1 \end{bmatrix} = \left[\begin{array}{c|c} \mathbf{Q}_S & \mathbf{R}_S \\ \hline \mathbf{0} & I \end{array} \right] \tag{8}$$

where p_S is the probability of occurrence of S ($= p_C + p_G$).

In similar way to the previous subsection the expected value of variable D_S is the $1/(p_S p_S)$. Under an independent symbol model the probability distribution of the CG distances can be written as a closed matrix form,

$$\begin{bmatrix} P(D_S = d_S | E_0) \\ P(D_S = d_S | E_1) \end{bmatrix} = \mathbf{Q}_S^{d-1} \begin{bmatrix} 0 \\ p_S \end{bmatrix}. \tag{9}$$

Naturally, for computing the inter SS distance distribution we must consider E_1 as the first state.

3   Experimental Procedure and Results

3.1   Materials

We analyze the human genome and use the reference assembly build 37.1 available from the website of the National Center for Biotechnology Information (http://www.ncbi.nlm.nih.gov/). All chromosomes of the human genome were processed as separate sequences.

We also use the Takai and Jones criterion to identify CpG island sequences and non CpG island sequences. We divided the chromosome sequences in 500 bp blocks, and we apply Takai and Jones criterion to determine if a block is a CpG island. In this work, we use this algorithm as the golden standard for finding CpG islands.

To evaluate performance on an independent data set, we used the set of CpG islands reported in Illingworth et al. [2, Supporting information]. We extracted each reported CpG island segment from the NCBI36 assembly build, release 54, unmasked human genomic sequences available from ftp://ftp.ensembl.org/.

3.2   Exploratory Analysis

In this work we analyzed the complete human genome, each chromosome, CpG islands and non CpG islands. We extracted the empirical inter CpG and SS distance sequence, we computed the corresponding distributions and we obtained the reference distribution under independent nucleotide (symbol) model.

Figure 3 shows the plot of the inter CpG distance distribution for the complete human genome and plot of the reference distribution (under independent symbol model). We can observe several differences between both distributions, but the most notable difference is that for distances lower than about 50 the

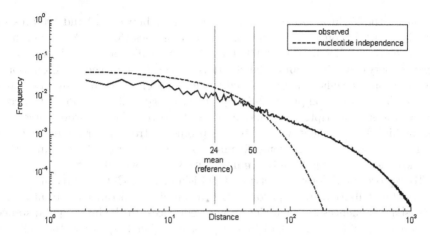

Fig. 3. Inter CpG distance distribution for the complete human genome and reference values.

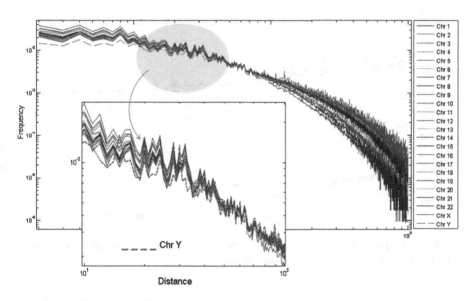

Fig. 4. Inter CpG distance distribution for chromosomes. The chromosome Y is highlighted.

empirical distances are under represented and for distances higher than about 50 the empirical distances are over represented.

To compare the inter CpG distances distribution profile between chromosomes, we used the empirical CpG distance distribution for each chromosome. Figure 4 presents all the chromosomes profiles together and we observe that there is a high correlation between profiles (min = 0.9395 and max = 0.9998). We

emphasize the Y chromosome because the distances between 10 and 100 present
a distinct profile compared to the other chromosomes (for the Y chromosome
the correlation varies between 0.9390 and 0.9778, while for other chromosomes
it varies between 0.9790 and 0.9998). It is known that in the Y chromosome
there are some specific features related to methylation which may be associated
with the Y chromosome profile differentiation. In particular, the lower CG con-
tent and less transcriptionally active genes. However, the X chromosome also
present high levels of methylation but its profile is strongly correlated with the
other chromosomes (chromosome X: min = 0.9790 and max = 0.9998; and other
chromosomes excluding X and Y: min = 0.9803 and max = 0.9998).

To explore the differences between CpG islands, non CpG islands and global
empirical distributions we also computed the empirical distance distribution for
CpG island sequences and non CpG island sequences. In the inter CpG distance
distribution profile the short distances present higher frequencies than all the
other regions under analysis (see Fig. 5).

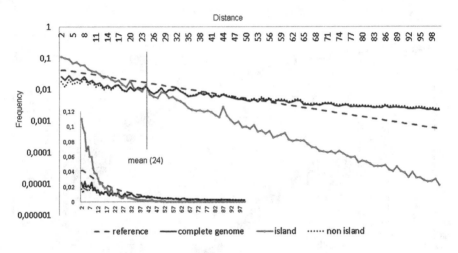

Fig. 5. Inter CpG distance distribution for the complete human genome, CpG islands,
non CpG islands and the reference distribution.

To compare the complete human genome, CpG islands and non CpG islands
we extract some statistics: the percentage of distances less than the reference
mean (24) and the median (17). Table 1 presents the results obtained and we
can observe different statistics values according to the type of sequence.

We also explored the inter CG distance distribution and the corresponding
reference distribution, the differences between the distance distributions of the
complete human genome and the reference are significant (p-values < 0.000, chi-
square test) but strongly correlated (correlation coefficient is 0.9988). Table 2
presents the results for some statistics in the SS distance context. The CpG
islands presented the most dissimilar statistic values.

Table 1. Some statistics for inter CpG distances for the reference, the complete human genome, the CpG islands and the non CpG islands in human genome.

	Reference	Human genome		
		Complete	CpG island	Non CpG island
Mean value	23.9	97.3	10.8	106.0
Median value	17	42	7	50
% < 8	23	13	50	1
% < 9	27	16	56	12
% < 10	30	18	61	13
% < reference median	49	27	80	22
% < reference mean	63	35	90	30

Table 2. Some statistics for inter SS distances for the reference, the complete human genome, the CpG islands and the non CpG islands in human genome.

	Reference	Human genome		
		Complete	CpG island	Non CpG island
Mean value	6.0	6.4	2.3	6.5
Median value	3	3	1	3
P_{80}	10	10	3	10
P_{90}	15	16	5	16
% < reference median	41	39	66	38
% < reference mean	62	65	91	64

From the observation of the results presented in Tables 1 and 2, we propose a procedure for CpG detection. We define a cutpoint (k_0) to categorize the CpG distances in long and short distances, the value of the median CpG distance in CpG islands can motivate the potential values for the cutpoint (Table 1). We define also two new criteria to identify CpG islands:

$$- \overline{d_S} < k_1;$$
$$- \frac{n(d < k_0)}{E[N(d < k_0)]} > k_2,$$

where $E[N(d < k_0)] = n_b * P(C) * P(G) * P(D < k_0)$ is the expected number of occurrences of short inter CpG distances in the block, and n_b is the length of the sequence block.

The first criterion (k_1) was motivated by the exploratory analysis of SS distances where the mean value in the CpG islands is much lower than the other sequences under study (Table 2). The second criterion was motivated by the exploratory analysis of CpG distances where the quantiles in the CpG islands are much higher than in the other sequences under study (Table 1).

We trained the criteria on chromosome 1 (human genome) and the values of k_i, $i = 0, 1, 2$, that lead to better results (in terms of accuracy, sensitivity and specificity) are:

- $k_0 \in \{7, 8, 9, 10\}$
- $k_1 \in [3.0; 5.0]$
- $k_2 \in [1.20; 2.39]$

In chromosome 1, a compromise solution was obtained for $k_0 = 9$, $k_1 = 4$ and $k_2 = 1.91$. We applied these criteria values to the other human chromosomes and the results obtained are shown in Table 3. The results reveal high accuracy, sensitivity and specificity. In order to compare our results with those of an independent algorithm, we also applied an algorithm based on HMM [2, chap. 3] to the same data. We analyzed the decoded state sequence for each 500 bp block, and classified the block as CpG island if more than 50 % of the states were island states. In the context of 500 bp block division, the global results using the inter-symbols distances are better than the HMM based algorithm.

Table 3. Performances, in relation to Takai and Jones criterion, of the proposed inter distances criterion, and of the HMM-based algorithm [2], measured for each chromosome of the human genome.

	Inter distances criterion			HMM with a posteriori probability cutoff of 0.5		
Chr	Accuracy (%)	Sensibility (%)	Specificity(%)	Accuracy (%)	Sensibility (%)	Specificity(%)
1	0.9863	0.9928	0.9862	0.9562	0.9909	0.9559
2	0.9911	0.9916	0.9911	0.9672	0.9846	0.9671
3	0.9922	0.9911	0.9923	0.9750	0.9900	0.9749
4	0.9924	0.9844	0.9924	0.9734	0.9860	0.9733
5	0.9912	0.9881	0.9912	0.9719	0.9891	0.9718
6	0.9909	0.9884	0.9909	0.9715	0.9878	0.9714
7	0.9854	0.9826	0.9854	0.9546	0.9825	0.9544
8	0.9901	0.9749	0.9902	0.9645	0.9701	0.9644
9	0.9848	0.9851	0.9848	0.9479	0.9862	0.9476
10	0.9884	0.9838	0.9884	0.9613	0.9822	0.9611
11	0.9862	0.9915	0.9862	0.9525	0.9904	0.9522
12	0.9866	0.9820	0.9866	0.9609	0.9814	0.9607
13	0.9916	0.9853	0.9916	0.9714	0.9789	0.9714
14	0.9876	0.9842	0.9878	0.9575	0.9873	0.9573
15	0.9879	0.9871	0.9879	0.9601	0.9906	0.9599
16	0.9723	0.9783	0.9722	0.9165	0.9771	0.9157
17	0.9704	0.9822	0.9702	0.9089	0.9861	0.9076
18	0.9910	0.9844	0.9911	0.9693	0.9831	0.9692
19	0.9417	0.9856	0.9404	0.8507	0.9865	0.8469
20	0.9815	0.9856	0.9815	0.9395	0.9864	0.9390
21	0.9831	0.9787	0.9831	0.9366	0.9748	0.9363
22	0.9685	0.9884	0.9681	0.8721	0.9865	0.8703
X	0.9909	0.9789	0.9910	0.9753	0.9901	0.9752
Y	0.9922	1.0000	0.9922	0.9786	0.9848	0.9786

We also evaluated the performance of the inter distances criterion using the CpG islands data set published by Illingworth et al. [9, Supporting information]. We applied three methods to the CpG island segments reported in the data set: the inter distances criterion, the Takai and Jones criterion and the HMM based algorithm [2]. Table 4 shows the success rates for all the methods used.

Table 4. CpG islands success rates in a set of experimentally obtained islands from [9] using the proposed algorithm (inter distances criterion); Takai and Jones criterion; and HMM [2].

	Criterion		
	Takai and Jones	Inter distances	HMM
Success rate	70.9 %	97.8 %	82.4 %

From the observation of the results shown in Table 4 we can conclude that for the set of experimentally obtained islands published in [9] the inter distances criterion has the best success rate.

4 Conclusion and Future Work

We found the reference distance distribution for CpG and SS dinucleotides (under independence nucleotide assumption). Almost all chromosomes present similar CpG distance profiles, but chromosome Y presents a distinct CpG distance profile. The CpG islands sequences present the highest frequencies of short distances. Based on inter CpG and inter SS distances, we proposed the inter distance criterion to classify DNA segments as CpG island (or not).

We conclude that the inter CpG and inter SS distances have the potential for contributing to the discrimination of CpG islands within DNA.

We expect that the inter CpG and inter SS symbol distances will be able to complement existing methods to increase the overall performance of CpG islands finding algorithms.

5 Funding

This work was supported by Portuguese funds through the CIDMA - Center for Research and Development in Mathematics and Applications, IEETA - Institute of Electronics and Telematics Engineering of Aveiro and the Portuguese Foundation for Science and Technology ("FCT–Fundação para a Ciência e a Tecnologia"), within projects PEst-OE/MAT/UI4106/2014 and PEst-OE/EEI/UI0127/2014. RMS is supported by the project Neuropath (CENTRO-07-ST24-FEDER-002034), co-funded by QREN "Mais Centro" program and the EU.

References

1. Deaton, A.M., Bird, A.: Cpg islands and the regulation of transcription. Genes Dev. **25**(10), 1010–1022 (2011)
2. Durbin, R., Eddy, S.R., Krogh, A., Mitchison, G.: Biological Sequence Analysis: Probabilistic Models of Proteins and Nucleic Acids. Cambridge University Press, Cambridge (1998)
3. Ehrlich, M.: DNA methylation in cancer: too much, but also too little. Oncogene **21**(35), 5400–5413 (2002)
4. Gardiner-Garden, M., Frommer, M.: Cpg islands in vertebrate genomes. J. Mol. Biol. **196**, 261–282 (1987)
5. Grinstead, C.M.: Introduction to Probability. American Mathematical Society, Washington, D.C. (1998)
6. Hackenberg, M., Previti, C., Luque-Escamilla, P.L., Carpena, P., Martinez-Aroza, J., Oliver, J.L.: CpGcluster: a distance-based algorithm for CpG-island detection. BMC Bioinformatics **7**, 446 (2006)
7. Hackenberg, M., Barturen, G., Carpena, P., Luque-Escamilla, P., Previti, C., Oliver, J.: Prediction of CpG-island function: CpG clustering vs. sliding-window methods. BMC Genomics **11**(1), 327 (2010)
8. Han, L., Zhao, Z.: CpG islands or CpG clusters: how to identify functional GC-rich regions in a genome? BMC Bioinformatics **10**, 65 (2009)
9. Illingworth, R., Kerr, A., DeSousa, D., Jäyrgensen, H., Ellis, P., Stalker, J., Jackson, D., Clee, C., Plumb, R., Rogers, J., Humphray, S., Cox, T., Langford, C., Bird, A.: A novel CpG island set identifies tissue-specific methylation at developmental gene loci. PLoS Biol. **6**(1), e22 (2008)
10. Takai, D., Jones, P.: The CpG island searcher: a new WWW resource. Silico Biol. **3**(3), 235–240 (2003)

The Alternating Least-Squares Algorithm for CDPCA

Eloísa Macedo$^{(\boxtimes)}$ and Adelaide Freitas

University of Aveiro, Aveiro, Portugal
{macedo,adelaide}@ua.pt

Abstract. Clustering and Disjoint Principal Component Analysis (CDP CA) is a constrained principal component analysis recently proposed for clustering of objects and partitioning of variables, simultaneously, which we have implemented in R language. In this paper, we deal in detail with the alternating least-squares algorithm for CDPCA and highlight its algebraic features for constructing both interpretable principal components and clusters of objects. Two applications are given to illustrate the capabilities of this new methodology.

Keywords: Principal Component Analysis · Clustering · K-means

1 Introduction

Principal Component Analysis (PCA) is a widely used tool in applied statistics for exploratory data analysis and dimensionality reduction. It has many important applications in different fields, such as neuroscience, computer graphics, image compression, meteorology, oceanography, and in gene expression [2].

In essence, PCA allows the reduction of the dimensionality of data by the detection of a lower number of uncorrelated variables, called components, that are able to explain the maximum variability of the data, i.e., the data compression is done with minimum information loss. An orthogonal transformation projects the data into a lower dimensional space along the directions where the data presents the highest variability. This statistical technique is useful to represent data by drawing a low-dimensional graph (e.g., in biplots) in order to find patterns hidden on data and to interpret relationships between samples and variables. PCA can be performed via singular value decomposition of the data matrix.

Since each principal component (PC) is a linear combination of all the original variables, i.e., with nonzero loadings, this can be considered a tremendous shortcoming for component interpretation. To overcome this difficulty, various PCA-based methodologies have been proposed in the recent years, for instance, based on rotation techniques or obtaining components with zero loadings. In this latter context, several major papers have been published. In [9], it is proposed a new methodology called Simple Principal Component Analysis, which idea is to restrict the components' loadings to be equal to −1, 0 or 1. In 2003, Jolliffe,

© Springer International Publishing Switzerland 2015
A. Plakhov et al. (Eds.): EmC-ONS 2014, CCIS 499, pp. 173–191, 2015.
DOI: 10.1007/978-3-319-20352-2_12

Trendafinov and Uddin [3] introduced SCoTLASS, which is a maximal variance approach that obtains components where a bound is introduced on the sum of the absolute values of the loadings, and some become zero. Later, in 2006, Zou, Hastie and Tibshirani [11] introduced the Sparse Principal Component Analysis, which aims to obtain modified principal components with sparse loadings. In [11], it is also proposed efficient algorithms to perform the new sparse PCA and some numerical experiments with real and simulated data are reported. In 2007, a new approach for sparse PCA via Semidefinite Programming was proposed in [1], based on a convex semidefinite relaxation of the sparse PCA problem. There are also reported numerical experiments for comparing that technique with others. More recently, in 2013, it is proposed in [4] a new sparse PCA and an iterative thresholding algorithm to estimate principal subspaces.

When dealing with real data sets, there may be the need of reducing not only the dimension of the variable space, but also to reveal some patterns among the objects. Obviously, this can be done by performing PCA on the variables and applying a clustering technique on the objects. The desirable scenario for data visualization and interpretation is to obtain non overlapping clusters of objects and disjoint or sparse principal components.

A new methodology called Clustering and Disjoint Principal Component Analysis (referred to hereafter as CDPCA) [8] was recently proposed for clustering of objects and partitioning of variables, simultaneously. It permits to cluster objects along a set of centroids and partition the variables into a reduced set of components, simultaneously, in order to maximize the between cluster deviance of the components in the reduced space. The CDPCA classification of data consists of the construction of groups based on the closeness and similarity among data.

In [8], the proposed CDPCA model is described as a joint model of K-means applied on the data matrix and PCA applied on the matrix of centroids. Hence, it depends on three parameter matrices: one matrix for allocating the objects into the clusters, one other for identifying the centroids and another one for identifying to the loading components. The least-squares estimators of these parameters can be obtained by solving a quadratic mixed continuous and integer optimization problem [8]. An alternating least-squares (ALS) algorithm based on four steps is suggested in [8] to solve the problem. Notice that the ALS algorithm can be considered as an heuristic that iteratively solves the optimization problem based on two basic steps: allocation of objects via K-means ([10]) and reduction of the variable space via application of PCA on the resulting centroids. In this paper, we describe a detailed two-step-based scheme of the ALS algorithm proposed in [8] for estimating the parameters of the CDPCA model. Unlike PCA, in CDPCA disjoint components are returned, and thus, each original variable contributes to a single component. It is worth mentioning that the obtained CDPCA score components may be correlated, unlike in PCA where uncorrelated components are provided.

Recently, we have implemented the CDPCA in a easy-to-use software application [5] using R language [6], which is available from the authors upon request.

Beside returning an assignment matrix for the allocation of objects into clusters and a component loading matrix which allows to allocate the variables into disjoint subsets, the main features of our R-based implementation of CDPCA include a plot of the data projected into the two dimensional space defined by the first two CDPCA components, and also a pseudo-confusion matrix when the real classification is known, permitting to summarize and visualize the (mis)classification of the objects. The goal of this paper is to explain and illustrate the algebraic features of the two essential steps in each iteration of the ALS algorithm. A toy example is included to show some transformations performed in each step of the ALS algorithm. Additionally, a numerical experiment using real data is presented. To execute these analyses we use our R-implemented function of CDPCA. Since the goal of this work is not focused on our R-based implementation, only brief reference to this function will be given in the numerical example.

The paper is organized as follows. Section 2 presents the theoretical background and tools needed for the CDPCA technique. Section 3 is devoted to highlight the algebraic features behind the CDPCA detailing the ALS algorithm step by step. In Sect. 4, application of CDPCA using data from a breast cancer study is presented and the results are compared with those obtained using PCA. Concluding remarks appear in Sect. 5.

2   The Methodology of CDPCA

In this section we describe the CDPCA, based on the paper [8].

2.1   Notation

First of all, let us introduce some notations and basic definitions that will be used throughout this work.

$\mathbf{X} = (x_{ij})$: Data matrix with I objects in rows and J variables in columns; $\mathbf{X}$ is assumed to be standardized.

P, Q: Desired number of clusters of objects and subsets of variables, respectively.

$\mathbf{U} = (u_{ip})$: Matrix defining an allocation of the I objects into P clusters; $\mathbf{U}$ is a $I \times P$ binary and row stochastic matrix defined as

$$\begin{cases} u_{ip} = 1, \text{ if the } i\text{-th object belongs to the cluster } p, \\ \\ u_{ip} = 0, \text{ otherwise.} \end{cases}$$

$\mathbf{V} = (v_{jq})$: Matrix defining a partition of the J variables into Q subsets; $\mathbf{V}$ is a $J \times Q$ binary and row stochastic matrix defined as

$$\begin{cases} v_{jq} = 1, \text{ if the } j\text{-th variable belongs to the subset } q, \\ \\ v_{jq} = 0, \text{ otherwise.} \end{cases}$$

$\bar{\mathbf{X}}$: Object centroid matrix in the original space; $\bar{\mathbf{X}}$ is a $P \times J$ matrix defined by $\bar{\mathbf{X}} = (\mathbf{U}^T\mathbf{U})^{-1}\mathbf{U}^T\mathbf{X}$.

$\mathbf{Z} = (z_{ij})$: Centroid-based data matrix where each object is identified by the corresponding centroid, i.e., each object is projected into the space defined by the P clusters; $\mathbf{Z}$ is a $I \times J$ matrix given by $\mathbf{Z} = \mathbf{U}\bar{\mathbf{X}}$.

$\mathbf{W}^{(q)} = \left(w_{ik}^{(q)}\right)$: Submatrix extracted from the centroid-based data matrix $\mathbf{Z}$ where only the original variables assigned into the q-th column of $\mathbf{V}$ are considered; $\mathbf{W}^{(q)}$ is a $I \times K^{(q)}$ matrix defined as

$$w_{ik}^{(q)} = z_{ij}, \text{ if } v_{jq} = 1, \quad \text{with } k = \text{rank}_{J^{(q)}}(j),$$

where $J^{(q)} = \{j : v_{jq} = 1\}$, $K^{(q)} = \#J^{(q)}$ and $k = 1, \cdots, K^{(q)}$.

$\mathbf{A} = (a_{jq})$: Matrix of the component loadings; $\mathbf{A}$ is a $J \times Q$ matrix where the Q columns are identifying the coefficients of Q linear combinations (i.e., the Q principal components for CDPCA) such that $\text{rank}(\mathbf{A}) = Q$, $\mathbf{A}^T\mathbf{A} = \mathbf{I}_Q$ and $\sum_{j=1}^{J}(a_{jq}a_{jr})^2 = 0$, for any q and r $(q \neq r)$.

$\mathbf{Y} = (y_{iq})$: Component score matrix where y_{iq} is the value of the i-th object for the q-th CDPCA component; $\mathbf{Y}$ is a $I \times Q$ matrix given by $\mathbf{Y} = \mathbf{X}\mathbf{A}$.

$\bar{\mathbf{Y}}$: Object centroid matrix in the reduced space; $\bar{\mathbf{Y}}$ is a $P \times Q$ matrix defined by $\bar{\mathbf{Y}} = \bar{\mathbf{X}}\mathbf{A}$.

2.2   Model

The CDPCA model results from the application of PCA on the transformed data matrix, where each object is replaced by its centroid. By its turn, the centroids are obtained by applying the K-means algorithm on the original data matrix [8].

Hence, the data matrix would be fitted by the model

$$\begin{aligned}
\mathbf{X} &= \mathbf{U}\bar{\mathbf{X}} + \mathbf{E}_1 & \text{(K-means applied on } \mathbf{X}) \\
&= \mathbf{U}\bar{\mathbf{Y}}\mathbf{A}^T + \mathbf{E}_1 + \mathbf{E}_2 & \text{(PCA applied on } \mathbf{U}\bar{\mathbf{X}}) \\
&= \mathbf{U}\bar{\mathbf{Y}}\mathbf{A}^T + \mathbf{E} & \text{(CDPCA model)}
\end{aligned} \qquad (1)$$

where $\mathbf{E}$, $\mathbf{E}_1$, $\mathbf{E}_2$ are $I \times J$ error matrices with $\mathbf{E} = \mathbf{E}_1 + \mathbf{E}_2$.

2.3   Optimization Problem

From the CDPCA model (1), it is easy to see that $\mathbf{E} = \mathbf{X} - \mathbf{U}\bar{\mathbf{Y}}\mathbf{A}^T$. Therefore, the CDPCA problem intents to minimize the norm of the error matrix $\mathbf{E}$, resulting in the following optimization problem

$$\min_{\mathbf{U},\bar{\mathbf{Y}},\mathbf{A}} \|\mathbf{X} - \mathbf{U}\bar{\mathbf{Y}}\mathbf{A}^T\|^2, \qquad (2)$$

subject to the above conditions for the matrices $\mathbf{U}$ (i.e., $\mathbf{U}$ is a binary and row stochastic matrix), $\bar{\mathbf{Y}}$ (i.e., $\bar{\mathbf{Y}}$ is an object centroid matrix in the reduced space) and $\mathbf{A}$ (i.e., $\mathbf{A}$ is a columnwise orthonormal matrix where each row contributes to a single column).

It can be proved that the problem (2) is equivalent to the maximization of the between cluster deviance $\|\mathbf{U}\bar{\mathbf{X}}\mathbf{A}\|^2$ of the components in the reduced space, subject to constraints on the matrices $\mathbf{U}$ and $\mathbf{A}$. Since the decomposition $\|\mathbf{X} - \mathbf{U}\bar{\mathbf{Y}}\mathbf{A}^T\|^2 = \|\mathbf{X}\|^2 - \|\mathbf{U}\bar{\mathbf{Y}}\mathbf{A}^T\|^2$ holds [8], the above problem (2) is equivalent to

$$\max_{\mathbf{U},\bar{\mathbf{Y}},\mathbf{A}} \|\mathbf{U}\bar{\mathbf{Y}}\mathbf{A}^T\|^2, \tag{3}$$

subject to the same constraints of problem (2). Since $\bar{\mathbf{Y}} = \bar{\mathbf{X}}\mathbf{A}$ and $\mathbf{A}$ has orthonormal columns (i.e., $\mathbf{A}^T\mathbf{A} = I$), then $\|\mathbf{U}\bar{\mathbf{Y}}\mathbf{A}^T\|^2 = \|\mathbf{U}\bar{\mathbf{X}}\mathbf{A}\|^2$. Hence, problem (3) is equivalent to

$$\max_{\mathbf{U},\bar{\mathbf{X}},\mathbf{A}} \|\mathbf{U}\bar{\mathbf{X}}\mathbf{A}\|^2. \tag{4}$$

To solve this optimization problem, the authors of CDPCA proposed the inclusion of the matrix $\mathbf{V}$ described in Sect. 2.1 which specifies the partition of J variables into Q disjoint components. The positions of the nonzero elements of the matrix $\mathbf{A}$ are identified by the positions of the one's in the matrix $\mathbf{V}$. Hence, and since $\bar{\mathbf{Y}} = \bar{\mathbf{X}}\mathbf{A}$, the CDPCA problem can be formulated as the following quadratic mixed continuous and integer problem:

$$
\begin{aligned}
\max \quad & F = \|\mathbf{U}\bar{\mathbf{Y}}\|^2 \\
\text{s. t.} \quad & u_{ip} \in \{0,\ 1\}, \quad i = 1,...,I;\ p = 1,...,P \\
& \sum_{p=1}^{P} u_{ip} = 1, \quad i = 1,...,I \\
& v_{jq} \in \{0,\ 1\}, \quad j = 1,...,J;\ q = 1,...,Q \\
& \sum_{q=1}^{Q} v_{jq} = 1, \quad j = 1,...,J \\
& \sum_{j=1}^{J} a_{jq}^2 = 1, \quad q = 1,...,Q \\
& \sum_{j=1}^{J} a_{jq}a_{jr} = 0, \quad q = 1,...,Q-1;\ r = q+1,...,Q
\end{aligned}
\tag{5}
$$

The first two constraints in (5) correspond to the allocation of I objects into P clusters. The following two constraints represent the allocation of J variables into Q disjoint subsets of variables (components). The remaining constraints are associated to the PCA implementation. The objective function value is calculated by $\|\mathbf{U}\bar{\mathbf{Y}}\|^2 = \operatorname{tr}\left(\mathbf{U}\bar{\mathbf{Y}}(\mathbf{U}\bar{\mathbf{Y}})^T\right)$, corresponding to the between cluster distances. Based on linear algebra properties, the objective function value F can also be equivalently computed by $\operatorname{tr}\left((\mathbf{U}\bar{\mathbf{Y}})^T\mathbf{U}\bar{\mathbf{Y}}\right)$, representing the total variance of the data in the reduced space, where the objects are identified by their centroids. The main goal is the achievement of maximum dissimilarity or distance between centroids (and objects) of different clusters. The idea of CDPCA is finding a clustering of objects along a set of centroids and, simultaneously, a partition of variables along a reduced set of disjoint components, in order to maximize the between cluster deviance in the reduced space of the disjoint components.

2.4   Algorithm

In [8], it is proposed an iterative algorithm called alternating least-squares algorithm (ALS) to solve the optimization problem (5). Each iteration of the ALS algorithm can be summarily described by two basic steps: allocation of objects via K-means and reduction of the variable space via application of PCA on the resulting centroids. Concretely,

- Step 1: Concerning to the objects:
  allocate the I objects into P clusters (matrix $\mathbf{U}$),
  calculate the centroids in the space of the observed variables (matrix $\bar{\mathbf{X}}$)
  identify the objects by its cluster centroids in the space of the observed variables (matrix $\mathbf{Z}$).
- Step 2: Concerning to the variables:
  allocate the J variables into Q subsets (matrix $\mathbf{V}$),
  obtain the loadings of the CDPCA components (matrix $\mathbf{A}$),
  calculate the centroids in the reduced space of the Q CDPCA components (matrix $\bar{\mathbf{Y}}$),
  identify the objects in the reduced space of the Q CDPCA components (matrix $\mathbf{Y}$).

These steps are summarized in Fig. 1. At the beginning, in Step 1 and with the standardized data matrix $\mathbf{X}$ of I objects described by J variables, the I objects are assigned into P clusters by means of the matrix $\mathbf{U}$. Next, each row of the data matrix is replaced by its corresponding object centroid resulting then in the matrix $\mathbf{Z}$. In Step 2, the allocation of the J variables into Q disjoint subsets is specified in the matrix $\mathbf{V}$ and the CDPCA component loadings are specified in the matrix $\mathbf{A}$. To obtain these two matrices, an iterative process working row-by-row and column-by-column of the matrices $\mathbf{V}$ and $\mathbf{A}$ is executed in order to maximize the objective function F. At the end of Step 2, the component score matrix, $\mathbf{Y}$, and the object centroid matrix in the reduced space, $\bar{\mathbf{Y}}$, are found as well as the value of the objective function F. Thus, at the end of one iteration of the algorithm, the I objects of the data matrix are allocated into P clusters, and simultaneously displayed in a reduced space of Q disjoint components. The value of the between cluster deviance is also calculated to evaluate the quality of the clustering of the I objects in the reduced space. In the next iteration, the process is repeated using $\mathbf{Y}$ as the input data matrix. The iterative procedure of the algorithm stops when there is a difference between consecutive computations of the values of the objective function F smaller than a specified tolerance.

Since the function F is bounded above, the algorithm converges to a stationary point, which is at least a local maximum of problem [8]. This procedure can be considered as an heuristic and thus, to guarantee that the global maximum is achieved, it has been suggested to run the algorithm several times for different initial allocation matrices $\mathbf{U}$ and $\mathbf{V}$, which are randomly chosen at the beginning of each run.

3   CDPCA: Step by Step

In this section, we present in detail the main algebraic features of the ALS algorithm for performing CDPCA.

To show the main algebraic features of the CDPCA procedure, we have performed CDPCA on a synthetic data matrix $\mathbf{X}$ constructed for satisfying the model (1) and where the objects are partitioned along a set of three clusters and the variables along a set of two components. For that purpose, we consider $I = 15$, $J = 3$, $P = 3$, $Q = 2$, and the following matrices satisfying the conditions mentioned in Sect. 2.1:

$$
\mathbf{U} = \begin{bmatrix} 1 & 0 & 0 \\ 1 & 0 & 0 \\ 1 & 0 & 0 \\ 1 & 0 & 0 \\ 1 & 0 & 0 \\ 0 & 1 & 0 \\ 0 & 1 & 0 \\ 0 & 1 & 0 \\ 0 & 1 & 0 \\ 0 & 0 & 1 \\ 0 & 0 & 1 \\ 0 & 0 & 1 \\ 0 & 0 & 1 \\ 0 & 0 & 1 \\ 0 & 0 & 1 \end{bmatrix}, \quad \bar{\mathbf{Y}} = \begin{bmatrix} \sqrt{2/3} & -2 \\ \sqrt{2/3} & 1 \\ -\sqrt{3/2} & 1 \end{bmatrix} \quad \text{and} \quad \mathbf{A} = \begin{bmatrix} 1 & 0 \\ 0 & \sqrt{2}/2 \\ 0 & \sqrt{2}/2 \end{bmatrix}. \tag{6}
$$

It is easy to check that, under these circumstances, $\mathbf{U}\bar{\mathbf{Y}}\mathbf{A}^T$ is a standardized matrix. An error $\mathbf{E}$ is added to obtain the model (1). Herein we considered the matrix $\mathbf{E}$ with values randomly generated of a normal distribution with mean zero and standard deviation equal to 0.8. Thus, we have

$$
\mathbf{X} = \mathbf{U}\bar{\mathbf{Y}}\mathbf{A}^T + \mathbf{E}
$$

being

$$
\mathbf{U}\bar{\mathbf{Y}}\mathbf{A}^T = \begin{bmatrix} 0.816 & -1.414 & -1.414 \\ 0.816 & -1.414 & -1.414 \\ 0.816 & -1.414 & -1.414 \\ 0.816 & -1.414 & -1.414 \\ 0.816 & -1.414 & -1.414 \\ 0.816 & 0.707 & 0.707 \\ 0.816 & 0.707 & 0.707 \\ 0.816 & 0.707 & 0.707 \\ 0.816 & 0.707 & 0.707 \\ -1.224 & 0.707 & 0.707 \\ -1.224 & 0.707 & 0.707 \\ -1.224 & 0.707 & 0.707 \\ -1.224 & 0.707 & 0.707 \\ -1.224 & 0.707 & 0.707 \\ -1.224 & 0.707 & 0.707 \end{bmatrix} \quad \mathbf{X} = \begin{bmatrix} 0.917 & -1.093 & -2.382 \\ 1.860 & -1.767 & -0.289 \\ 0.460 & -1.509 & -0.132 \\ 1.290 & -0.412 & -1.405 \\ 1.567 & -1.812 & -1.530 \\ 0.982 & -0.167 & 1.531 \\ 0.832 & 1.710 & 0.334 \\ 2.461 & 1.315 & 1.203 \\ 0.697 & 1.520 & 1.519 \\ -2.273 & 0.152 & 0.464 \\ -1.603 & 1.483 & -0.476 \\ -1.003 & -0.043 & -0.840 \\ -0.799 & 1.763 & -0.770 \\ -1.133 & 0.002 & 1.145 \\ -2.599 & 0.473 & 0.393 \end{bmatrix}
$$

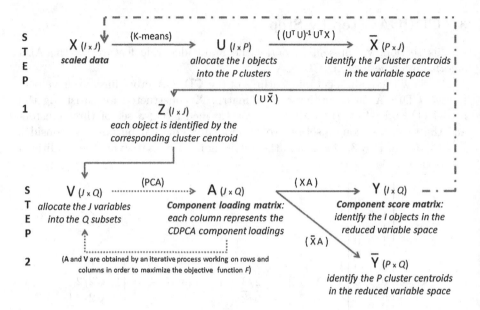

Fig. 1. The two basic steps of one iteration of the ALS algorithm for performing CDPCA.

The dashed horizontal and vertical lines separate the three clusters of objects and the set of variables, respectively, and in accordance with (6).

Using the synthetic data matrix **X**, we now focus on the algebraic features behind the two basic steps of the CDPCA methodology and afterwards illustrate some outputs obtained by our R-based application.

3.1 Initialization

Set $k = 0$. At the beginning, the data matrix **X** is standardized:

$$x_{ij} \mapsto \frac{x_{ij} - \bar{x}_j}{\sqrt{\sum_{i=1}^{I} (x_{ij} - \bar{x}_j)^2 / I}}$$

where $\bar{x}_j = \sum_{i=1}^{I} x_{ij}/I$. Next, the parameters of the ALS algorithm to perform CDPCA are initialized as follows:

Step 1. Parameters associated to the objects:
- The matrix $\mathbf{U}_0$ is randomly generated such that there is only a nonzero element per row and that element is equal to 1 (i.e., $\mathbf{U}_0$ is the initial object assignment matrix).
- The object centroid matrix $\bar{\mathbf{X}}_0$ is computed. For such, the mean of each variable into each object cluster is calculated.

- All the objects are identified by its cluster centroids. This information is provided by the centroid-based data matrix $\mathbf{Z}$.

Step 2. Parameters associated to the components:
- The matrix $\mathbf{V}_0$ is randomly generated such that there is only a nonzero element per row and that element is equal to 1 (i.e., $\mathbf{V}_0$ is the initial variable assignment matrix).
- The CDPCA component loading matrix $\mathbf{A}_0$ is constructed column-by-column solving Q independent PCA subproblems, one by each column. The nonzero elements of the q-th column of $\mathbf{V}_0$ identify the original variables belonging to the q-th CDPCA component. These elements will be considered in the PCA subproblem to obtain the nonzero elements of the q-th column of $\mathbf{A}_0$. Thus, the nonzero elements on the q-th column of $\mathbf{A}_0$ correspond to the first principal component obtained from PCA applied on the submatrix $\mathbf{W}_0^{(q)}$ which is extracted from the centroid-based data matrix $\mathbf{Z}_0 = \mathbf{U}_0 \bar{\mathbf{X}}_0$ (i.e., the data matrix where each object is identified by the corresponding centroid) and restricted to the original variables assigned into the q-th column of $\mathbf{V}_0$. Therefore, the q-th column of $\mathbf{A}_0$ provides the direction vector with maximum variability among the centroids in the subspace defined by the original variables assigned to the q-th column of $\mathbf{V}_0$.

3.2 General Iteration

At the beginning of the $(k + 1)$-th iteration of the algorithm, the matrices $\mathbf{U}_k$, $\bar{\mathbf{X}}_k$ $\mathbf{V}_k$, $\mathbf{A}_k$ and $\bar{\mathbf{Y}}_k$ are known.

Step 1. Parameters associated to the objects:
The matrix $\mathbf{U}_{k+1}$ is given by one run of the K-means algorithm on the score matrix $\mathbf{Y}_k = \mathbf{X}\mathbf{A}_k$ starting from the object centroid matrix $\bar{\mathbf{Y}}_k$ in the reduced space. The P new clusters are obtained finding the new centroids, i.e., updating the centroid matrix by $\bar{\mathbf{X}}_{k+1} = \left(\mathbf{U}_{k+1}^T \mathbf{U}_{k+1}\right)^{-1} \mathbf{U}_{k+1}^T \mathbf{X}$ and the object centroid-based matrix by $\mathbf{Z}_{k+1} = \mathbf{U}_k \bar{\mathbf{X}}_k$.

Every cluster should be assigned with at least one object. On the procedure, if any cluster becomes empty, then a selection step is fulfilled: half of the objects on the bigger cluster is assigned into one of the empty clusters, and this process is repeated while there are empty clusters.

Step 2. Parameters associated to the components:
The updated matrices $\mathbf{V}_{k+1}$ and $\mathbf{A}_{k+1}$ are sequentially constructed row-by-row, and in each row, the process is also sequentially performed column-by-column, in a symbiotic relationship with the maximization of the objective function F.

The matrix $\mathbf{V}$ specifies a partition of the original variables into Q disjoint components. For updating $\mathbf{V}_k$, each original variable will be evaluated in order to find which component leads to a higher value of the objective function F, assuming that all remaining variables are fixed in the components in accordance with $\mathbf{V}_k$.

Firstly, the first row of $\mathbf{V}_k$ is updated by detecting for which column j, with $j = 1, \cdots, Q$, the allocation of its nonzero element yields better results in the sense of the maximization of the objective function. Concretely, for the first row (variable) of $\mathbf{V}_{k+1}$, the *best* column (component) among Q is selected by solving Q PCA subproblems associated to the updated matrices $\mathbf{W}_{k+1}^{(q)}$, for $q = 1, 2, \cdots, Q$, respectively, assuming the Q possible positions of the nonzero element into the first row of the potential updated matrix $\mathbf{V}_{k+1}$. In the q-th PCA subproblem, the first principal component is calculated determining the update of the q-th column of $\mathbf{A}_{k+1}$. At this point, the centroid matrix on the reduced space, $\bar{\mathbf{Y}}_{k+1}$, and the objective function value, F_{k+1}, can be computed by $\bar{\mathbf{Y}}_{k+1} = \bar{\mathbf{X}}_{k+1}\mathbf{A}_{k+1}$ and $F_{k+1} = \text{tr}\left((\mathbf{U}_{k+1}\bar{\mathbf{Y}}_{k+1})^T\mathbf{U}_{k+1}\bar{\mathbf{Y}}_{k+1}\right)$. This process is done repeatedly to select the *best* component to allocate the first row (variable) in $\mathbf{V}_{k+1}$, which will coincide with the component that yields the highest value of F_{k+1}.

The same process is now repeated for the remaining rows of $\mathbf{V}_k$, and therefore, $\mathbf{V}_{k+1}$ is updated row-by-row. Hence, for each original variable there are solved Q assignment subproblems. In each subproblem, a subspace of variables is considered and the best direction (eigenvector) with maximum variability explained is obtained performing a PCA step. Each variable will be included into a component associated to the subproblem that maximizes the objective function.

Since there are J original variables, i.e., J rows on $\mathbf{V}_k$, then there are $J \times Q$ subproblems to be solved in order to obtain $\mathbf{V}_{k+1}$ and $\mathbf{A}_{k+1}$. At the end of the Step 2, the best assignment will maximize the objective function, and consequently, the between cluster deviance given by $F_{k+1}/\|\mathbf{Y}_{k+1}\|^2$, where $\mathbf{Y}_{k+1} = \mathbf{X}\mathbf{A}_{k+1}$.

Stopping Criterion. Evaluate solutions:

If the difference between F_k and F_{k+1} is smaller than a specified tolerance, then the algorithm stops and returns the current iterates. Otherwise, repeat the iteration, setting $k := k + 1$.

At the end of the algorithm, say, for instance, at the k^*-th iteration, besides returning the allocation matrices $\mathbf{U}_{k^*}$, for the objects, and $\mathbf{V}_{k^*}$, for the variables, the component loading matrix $\mathbf{A}_{k^*}$ is also returned, which is a columnwise orthonormal matrix whose elements are the loadings of the CDPCA components. Moreover, the CDPCA component score matrix $\mathbf{Y}_{k^*}$ is obtained, as well as the object centroid matrix in the reduced space $\bar{\mathbf{Y}}_{k^*}$. These matrices can be used to obtain an approximation of the CDPCA model by

$$\mathbf{U}_{k^*}\bar{\mathbf{Y}}_{k^*}\mathbf{A}_{k^*}^T,$$

providing a partition of the objects along a set of clusters and the variables along a set of disjoint components.

It is worth mentioning that, unlike in the PCA technique, the ALS algorithm can not establish the CDPCA components decreasingly sorted by their explained variability. In order to be consistent with the classical form of representation of the components, at the end of the algorithm the columns of the matrices associated to the CDPCA components, namely, $\mathbf{V}_{k*}$, $\mathbf{A}_{k*}$, $\mathbf{Y}_{k*}$, and $\bar{\mathbf{Y}}_{k*}$, will be rearranged. Since the changes are performed in all of these matrices, the above CDPCA model is trivially satisfied with the rearranged matrices.

3.3   Synthetic Data

In the following we illustrate an execution of the ALS algorithm described above using the synthetic data. The data matrix is formed by $I = 15$ objects and $J = 3$ variables. In order to evaluate the performance of the algorithm we will also analyse the ability of the algorithm for detecting the $P = 3$ clusters of objects and $Q = 2$ subsets of variables known in the synthetic data.

Considering the synthetic data, we set $k = 0$, specify the convergence tolerance as $\varepsilon = 10^{-5}$, and initialize the parameters of the CDPCA model.

Initialization:
In the Step 1, we get

$$
\mathbf{U}_0 = \begin{bmatrix} 1 & 0 & 0 \\ 0 & 1 & 0 \\ 0 & 0 & 1 \\ 1 & 0 & 0 \\ 1 & 0 & 0 \\ 0 & 1 & 0 \\ 1 & 0 & 0 \\ 0 & 1 & 0 \\ 0 & 0 & 1 \\ 0 & 0 & 1 \\ 0 & 0 & 1 \\ 0 & 0 & 1 \\ 0 & 1 & 0 \\ 0 & 0 & 1 \\ 0 & 0 & 1 \end{bmatrix}, \
\bar{\mathbf{X}}_0 = \begin{bmatrix} 0.690 & -0.416 & -1.022 \\ 0.673 & 0.145 & 0.440 \\ -0.779 & 0.154 & 0.332 \end{bmatrix}, \
\mathbf{Z}_0 = \begin{bmatrix} 0.690 & -0.416 & -1.022 \\ 0.673 & 0.145 & 0.440 \\ -0.779 & 0.154 & 0.332 \\ 0.690 & -0.416 & -1.022 \\ 0.690 & -0.416 & -1.022 \\ 0.673 & 0.145 & 0.440 \\ 0.690 & -0.416 & -1.022 \\ 0.673 & 0.145 & 0.440 \\ -0.779 & 0.154 & 0.332 \\ -0.779 & 0.154 & 0.332 \\ -0.779 & 0.154 & 0.332 \\ -0.779 & 0.154 & 0.332 \\ 0.673 & 0.145 & 0.440 \\ -0.779 & 0.154 & 0.332 \\ -0.779 & 0.154 & 0.332 \end{bmatrix}
$$

In the Step 2, it begins with

$$
\mathbf{V}_0 = \begin{bmatrix} 1 & 0 \\ 0 & 1 \\ 1 & 0 \end{bmatrix}.
$$

Next, we determine $\mathbf{A}_0$. Fixing the first column of $\mathbf{V}_0$ ($q = 1$), the unit normed eigenvector $v_0^{(1)}$ associated to the largest eigenvalue of the correlation matrix of the submatrix $\mathbf{W}_0^{(1)}$ is selected and introduced in the nonzero entries of the first column of $\mathbf{A}_0$. A similar procedure is performed for the remaining columns of $\mathbf{V}_0$. Thus, for $q = 1$, $K^{(1)} = 2$ and we get the 15×2 matrix

$$\mathbf{W}_0^{(1)} = \begin{bmatrix} 0.690 & -1.022 \\ 0.673 & 0.440 \\ -0.779 & 0.332 \\ 0.690 & -1.022 \\ 0.690 & -1.022 \\ 0.673 & 0.440 \\ 0.690 & -1.022 \\ 0.673 & 0.440 \\ -0.779 & 0.332 \\ -0.779 & 0.332 \\ -0.779 & 0.332 \\ -0.779 & 0.332 \\ 0.673 & 0.440 \\ -0.779 & 0.332 \\ -0.779 & 0.332 \end{bmatrix}.$$

The unit normed eigenvector associated to the largest eigenvalue of the 2×2 matrix $\left(\mathbf{W}_0^{(1)}\right)^T \mathbf{W}_0^{(1)}$ is given by $v_0^{(1)} = \begin{bmatrix} -0.809 \\ 0.587 \end{bmatrix}$. Hence, for the nonzero elements on the first column of $\mathbf{A}_0$, which correspond to the nonzero entries on the first column of $\mathbf{V}_0$, we shall introduce $v_0^{(1)}$. Similarly, considering now the second column of $\mathbf{V}_0$, we have $q = 2$, $K^{(2)} = 1$ and $\left(\mathbf{W}_0^{(2)}\right)^T \mathbf{W}_0^{(2)}$ is a 1×1 matrix. Thus, we get

$$\mathbf{W}_0^{(2)} = \begin{bmatrix} -0.416 \\ 0.145 \\ 0.154 \\ -0.416 \\ -0.416 \\ 0.145 \\ -0.416 \\ 0.145 \\ 0.154 \\ 0.154 \\ 0.154 \\ 0.154 \\ 0.145 \\ 0.154 \\ 0.154 \end{bmatrix} \qquad \text{and} \qquad v_0^{(2)} = [1],$$

and the nonzero element on the second column of $\mathbf{A}_0$ will be 1. Therefore, the CDPCA component loading matrix is given by

$$\mathbf{A}_0 = \begin{bmatrix} -0.809 & 0 \\ 0 & 1 \\ 0.587 & 0 \end{bmatrix}.$$

At this point, the objects of the data matrix $\mathbf{X}$ can be assigned in the reduced space of the CDPCA components by the object centroid matrix in the reduced space, $\bar{\mathbf{Y}}_0$, and the objective function F should be evaluated for the current matrices $\mathbf{U}_0$ and $\bar{\mathbf{Y}}_0$. Regarding our example, $\mathbf{Y}_0$ is a 15×2 matrix and $\bar{\mathbf{Y}}_0$ is a 3×2 matrix given as follows.

$$\mathbf{Y}_0 = \begin{bmatrix} -1.621 & -0.981 \\ -1.046 & -1.533 \\ -0.213 & -1.322 \\ -1.316 & -0.425 \\ -1.529 & -1.570 \\ 0.365 & -0.225 \\ -0.172 & 1.310 \\ -0.598 & 0.987 \\ 0.512 & 1.154 \\ 1.561 & 0.036 \\ 0.716 & 1.124 \\ 0.206 & -0.123 \\ 0.132 & 1.353 \\ 1.302 & -0.085 \\ 1.700 & 0.298 \end{bmatrix}, \qquad \bar{\mathbf{Y}}_0 = \begin{bmatrix} -1.159 & -0.416 \\ -0.286 & 0.145 \\ 0.826 & 0.154 \end{bmatrix}.$$

initial approximation of $\mathbf{Y}_0$ provides a partition of objects along a set of three clusters (objects 1, 4, 5 and 7 are currently assigned into one cluster, objects 2, 6, 8 and 13 are assigned into another cluster, and the remaining objects are currently belonging to a third cluster) and also a partition of variables along a set of disjoint components ($PC_1 = -0.809X_1 + 0.587X_3$ and $PC_2 = X_2$). Notice that the current partition does not correspond to the final solution, nor to the real partition; this is the result after computing the initial step of the CDPCA procedure. Additionally, the objective function value for the current iterates is $F_0 = 11.438$ and the corresponding between cluster deviance is $F_0/\|\mathbf{Y}_0\|_2^2 = 36.63\%$.

First iteration:
Set $k = 1$. In Step 1, the matrix of the allocation of objects into P clusters and the object centroid matrix are updated yielding the following matrices:

$$\mathbf{U}_1 = \begin{bmatrix} 1 & 0 & 0 \\ 1 & 0 & 0 \\ 1 & 0 & 0 \\ 1 & 0 & 0 \\ 1 & 0 & 0 \\ 0 & 0 & 1 \\ 0 & 1 & 0 \\ 0 & 1 & 0 \\ 0 & 0 & 1 \\ 0 & 0 & 1 \\ 0 & 0 & 1 \\ 0 & 1 & 0 \\ 0 & 1 & 0 \\ 0 & 0 & 1 \\ 0 & 0 & 1 \end{bmatrix}, \quad \bar{\mathbf{X}}_1 = \begin{bmatrix} 0.735 & -1.166 & -0.936 \\ 0.174 & 0.882 & 0.056 \\ -0.728 & 0.384 & 0.743 \end{bmatrix}, \quad \mathbf{Z}_1 = \begin{bmatrix} 0.735 & -1.166 & -0.936 \\ 0.735 & -1.166 & -0.936 \\ 0.735 & -1.166 & -0.936 \\ 0.735 & -1.166 & -0.936 \\ 0.735 & -1.166 & -0.936 \\ -0.728 & 0.384 & 0.743 \\ 0.174 & 0.882 & 0.056 \\ 0.174 & 0.882 & 0.056 \\ -0.728 & 0.384 & 0.743 \\ -0.728 & 0.384 & 0.743 \\ -0.728 & 0.384 & 0.743 \\ 0.174 & 0.882 & 0.056 \\ 0.174 & 0.882 & 0.056 \\ -0.728 & 0.384 & 0.743 \\ -0.728 & 0.384 & 0.743 \end{bmatrix}$$

Notice that $\mathbf{U}_1$ specifies a new allocation of the objects.

In Step 2, we get

$$\mathbf{V}_1 = \begin{bmatrix} 1 & 0 \\ 0 & 1 \\ 1 & 0 \end{bmatrix}, \quad \mathbf{A}_1 = \begin{bmatrix} -0.660 & 0 \\ 0 & 1 \\ 0.750 & 0 \end{bmatrix},$$

$$\mathbf{Y}_1 = \begin{bmatrix} -1.870 & -0.981 \\ -0.903 & -1.533 \\ -0.186 & -1.322 \\ -1.389 & -0.425 \\ -1.593 & -1.570 \\ 0.682 & -0.225 \\ -0.041 & 1.310 \\ -0.182 & 0.987 \\ 0.799 & 1.154 \\ 1.405 & 0.036 \\ 0.491 & 1.124 \\ -0.011 & -0.123 \\ -0.055 & 1.353 \\ 1.355 & -0.085 \\ 1.502 & 0.298 \end{bmatrix} \quad \text{and} \quad \bar{\mathbf{Y}}_1 = \begin{bmatrix} -1.188 & -1.166 \\ -0.072 & 0.882 \\ 1.039 & 0.384 \end{bmatrix}.$$

At the end of the first iteration, $F_1 = 24.374$ and the corresponding between cluster deviance is 77.93 %. The following step is to check the stopping criterion. Since $|F_1 - F_0| = 12.936 > \varepsilon$, another iteration should be computed.

Further iterations:

In order to refine the solutions, more iterations of the algorithm are needed. In this example, the best solution was obtained after two iterations and it took only 0.0 s to exhibit a solution. The obtained results are as follows. The object allocation matrix $\mathbf{U}$ and the variable allocation matrix $\mathbf{V}$, the component loading matrix $\mathbf{A}$, the component score matrix $\mathbf{Y}$ and the centroid matrix in the reduced space $\bar{\mathbf{Y}}$ already rearranged by column (in decreasing order of the variability explained by the CDPCA) are given by

$$\mathbf{U} = \begin{bmatrix} 1 & 0 & 0 \\ 1 & 0 & 0 \\ 1 & 0 & 0 \\ 1 & 0 & 0 \\ 1 & 0 & 0 \\ 0 & 1 & 0 \\ 0 & 1 & 0 \\ 0 & 1 & 0 \\ 0 & 1 & 0 \\ 0 & 0 & 1 \\ 0 & 0 & 1 \\ 0 & 0 & 1 \\ 0 & 0 & 1 \\ 0 & 0 & 1 \\ 0 & 0 & 1 \end{bmatrix}, \quad \mathbf{V} = \begin{bmatrix} 0 & 1 \\ 1 & 0 \\ 1 & 0 \end{bmatrix},$$

$$\mathbf{A} = \begin{bmatrix} 0 & 1 \\ -0.734 & 0 \\ -0.678 & 0 \end{bmatrix}, \quad \mathbf{Y} = \begin{bmatrix} 2.092 & 0.535 \\ 1.250 & 1.160 \\ 1.001 & 0.232 \\ 1.101 & 0.782 \\ 2.017 & 0.966 \\ -0.796 & 0.578 \\ -1.211 & 0.478 \\ -1.492 & 1.559 \\ -1.803 & 0.389 \\ -0.352 & -1.581 \\ -0.591 & -1.137 \\ 0.542 & -0.738 \\ -0.584 & -0.603 \\ -0.669 & -0.825 \\ -0.503 & -1.797 \end{bmatrix}.$$

$$\bar{\mathbf{Y}} = \begin{bmatrix} 1.492 & 0.735 \\ -1.325 & 0.751 \\ -0.359 & -1.113 \end{bmatrix},$$

The maximum for the objective function is 31.357 and the corresponding between cluster deviance is 85.63 %.

Our R-based implementation of this new methodology provides the graphical display of the CDPCA classification taking the first two CDPCA components, as well as the real classification when it is known. For the synthetic data, the plot is displayed in Fig. 2.

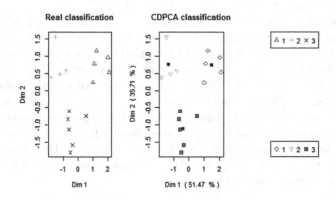

Fig. 2. Real and CDPCA classification for the synthetic data.

Clearly, the CDPCA was able to fulfil the classification and the objects were correctly assigned to the clusters.

Besides that, our R function also returns a pseudo-confusion matrix, here displayed in Table 1. The pseudo-confusion matrix allows one to easily verify how many objects are correctly assigned into clusters, or how many objects are misclassified.

Table 1. Pseudo-confusion matrix for the synthetic data set.

Real class	CDPCA class		
	1	2	3
1	5	0	0
2	0	4	0
3	0	0	6

From Table 1, we can observe that 5 objects are assigned to a cluster, 4 objects are assigned into a second cluster and the remaining 6 objects belong to another cluster. This table confirms the high accuracy classification produced by CDPCA on the Synthetic data.

4   Numerical Experiments

Here, we describe the numerical experiments of the CDPCA applied on a real data set. Our experiments were run on a computer with an Intel Core i5-3317U CPU @ 1.70 GHz, with Windows 7 (64 bits) and 6 GB RAM, using R version 3.0.0 (2013).

The CDPCA was implemented in R under the function CDpca [5]. This function is suitable for data matrices of numeric values.

Since the ALS algorithm can be considered as a heuristic, it is advisable to run the algorithm several times, as it has been suggested in [8], in order to find the global maximum. Therefore, all the presented numerical tests were run 1000 times and the tolerance for convergence purposes was set to 10^{-5}.

Our R implementation of CDPCA starts by standardizing the data. Among other outputs, the CDpca function returns the CDPCA component loading matrix, the obtained between cluster deviance, the objects assignment matrix, the variables assignment matrix, a pseudo-confusion matrix when the real classification is known a priori, the variance explained by the CDPCA components and a plot of the data projected into the two dimensional space defined by the first two components is displayed.

4.1   Breast Cancer Data

The Wisconsin Breast Cancer Database [7] contains 683 instances (originally, there were 699 instances; however, 16 of them were excluded since they contain missing values), where each of them is described by 9 attributes with integer values in the range $1-10$ and a real binary class label, which divides the instances into two classes: benign or malignant. The list of variables is formed by clump thickness, uniformity of cell size, uniformity of cell shape, marginal adhesion, single epithelial cell size, bare nuclei, bland chromatin, normal nucleoli, and the ninth variable mitoses describes an analysis of mitotic stages. These variables are used in pathology reports for suggesting whether a lump in a breast is benign or malignant.

The CDPCA was applied in this data set, by choosing $P = 2$ clusters of objects and $Q = 2$ subsets of variables and executing our CDpca function in R. It took only 6 iterations and 0.19 s to yield a solution approximation satisfying the convergence tolerance. The results of CDPCA are displayed in Tables 2, 3 and Fig. 3.

The Table 2 reports the component loadings for both PCA and CDPCA.

Comparing the results in Table 2, performing an analysis of data from the obtained results using the PCA technique can be complex. The resulting PCA component loadings lead to components which do not seem interpretable. This is due to all the original variables contribute to both PCA components and, therefore, it is quite difficult to detect a pattern or relation among the variables for each of the two first principal components. With CDPCA the interpretation of the components becomes easier, since each variable contributes to a single component.

Table 2. Component loadings for PCA and CDPCA on the Breast Cancer Data.

Variables	PCA loadings		CDPCA loadings	
	Component 1	Component 2	Component 1	Component 2
Clump Thickness	−0.296	−0.073	0.350	0
Uniformity of Cell Size	−0.403	0.229	0.429	0
Uniformity of Cell Shape	−0.392	0.164	0.426	0
Marginal Adhesion	−0.331	−0.098	0	−0.710
Single Epithelial Cell Size	−0.249	0.200	0	−0.703
Bare Nuclei	−0.442	−0.780	0.415	0
Bland Chromatin	−0.292	0.008	0.387	0
Normal Nucleoli	−0.354	0.469	0.374	0
Mitoses	−0.124	0.188	0.216	0
Explained variance (%)	69.05	7.20	51.71	17.74

The first PCA component explains $69,05\%$ of the total variance and is mainly characterized by Bare Nuclei, Uniformity of Cell Size and Uniformity of Cell Shape, while the second PCA component explains only $7,20\%$ of the total variance and is mainly characterized by Bare Nuclei and Normal Nucleoli. Notice that the variable Bare Nuclei is the most contributing variable for both components.

Considering now the CDPCA technique, it can be observed that the first CDPCA component explains $51,71\%$ of the total variance and is mainly characterized by Uniformity of Cell Size, Uniformity of Cell Shape and Bare Nuclei, while the second CDPCA component is only characterized by the original variables Marginal Adhesion and Single Epithelial Cell Size, explaining $17,74\%$ of the total variance.

Table 3. Pseudo-confusion matrix for the Breast Cancer data.

Real class	Preditive CDPCA class	
	1	2
	(benign)	(malignant)
1 (benign)	434	10
2 (malignant)	19	220

Table 3 evaluates the predictive performance of CDPCA as a classification technique on the Breast Cancer data. The real classification for this data set is as follows: 444 objects into the benign class, and 239 into the malignant. Considering the pseudo-confusion matrix obtained with the results on the CDPCA classification, we conclude that 453 objects are assigned to the benign class and 230 are included into the malignant class. This means that there are 29 misclassified objects, leading to a 4% of misclassification. Therefore, the CDPCA

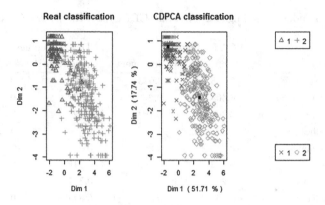

Fig. 3. Real and CDPCA classification for the Breast Cancer data.

classification presents an accuracy of 96 % permitting to conclude that our implementation of the CDPCA performed very well in practice.

In Fig. 3, a graph representation of the data into the 2-dimensional reduced space defined by the first two CDPCA components is depicted. This graph permits to visualize the data in order to help on the detection of patterns hidden in the data set. In the case of the Breast Cancer data, the graph shows that positive value for the first CDPCA component is tendentiously attributed to subjects (objects) with malignant lumps (class 2).

The obtained CDPCA between cluster deviance is 80,20 % of the total deviance.

5   Conclusions

Applications of the recently developed methodology CDPCA to data reveal that this method can be successful for classifying the samples and exploring relationship between variables, as well as for visualizing data into a reduced space. This paper is particularly focussed on detailing a two-step-based scheme of the ALS algorithm used to perform CDPCA and on its algebraic features. A toy example is included to illustrate the resulting transformations on the ALS algorithm step by step. A final remark is that the ALS algorithm for CDPCA performed very well and also revealed high accuracy in the clusterings for the presented examples and several other not shown herein.

Acknowledgments. The authors would like to thank the anonymous referee for all the valuable and constructive comments which have helped to improve this paper. A special thanks to Professor Maurizio Vichi for providing us a Matlab version of the ALS algorithm for performing CDPCA. This work was partially supported by Portuguese funds through the CIDMA - Center for Research and Development in Mathematics and Applications, and the Portuguese Foundation for Science and Technology (FCT – Fundação para a Ciência e a Tecnologia), within project UID/MAT/04106/2013.

References

1. d'Aspremont, A., El Ghaoui, L., Jordan, M.I., Lanckriet, G.R.G.: A direct formulation for sparse PCA using semidefinite programming. SIAM **49**(3), 434–448 (2007)
2. Jolliffe, I.T.: Principal Component Analysis, 2nd edn. Springer, New York (2002)
3. Jolliffe, I.T., Trendafilov, N.T., Uddin, M.: A modified principal component technique based on the lasso. J. Comput. Graph. Stat. **12**(3), 531–547 (2003)
4. Ma, Z.: Sparse principal component analysis and iterative thresholding. Ann. Stat. **41**(2), 772–801 (2013)
5. Macedo, E., Freitas, A.: Statistical methods and optimization in data mining. In: III International Conference of Optimization and Applications, OPTIMA 2012, pp. 164–169 (2012)
6. R Development Core Team: R: A Language and Environment for Statistical Computing. R Foundation for Statistical Computing. http://www.R-project.org/
7. UCI Repository: Winsconsin Breast Cancer Data Set. http://archive.ics.uci.edu/ml/datasets/Breast+Cancer+Wisconsin+(Original)
8. Vichi, M., Saporta, G.: Clustering and disjoint principal component analysis. Comput. Stat. Data Anal. **53**, 3194–3208 (2009)
9. Vines, S.: Simple principal components. Appl. Stat. **49**, 441–451 (2000)
10. Xu, R., Wunsch, D.: Survey of clustering algorithms. IEEE Trans. Neural Netw. **16**, 645–648 (2005)
11. Zou, H., Hastie, T., Tibshirani, R.: Sparse principal component analysis. J. Comput. Graph. Stat. **15**(2), 262–286 (2006)

Rapid Spatial Aggregation

Markus Loecher$^{(\boxtimes)}$ and Madhav Kumar

Berlin School of Economics and Law, Berlin, Germany
mloecher@hwr-berlin.de

Abstract. Data visualization is an important component of spatial data analysis. We demonstrate the visualization of spatial/spatio-temporal data on map tiles as implemented in the R package *RgoogleMaps*. We argue that extremely large spatial or location data sets can lead to clutter and information overload necessitating aggregation to higher geographical identities. Such aggregation requires associating each coordinate point from the set to a particular spatial polygon in the search space. Examples for such polygon-based spatial partitions would be zip codes, census blocks, or school districts. Unless efficient data structures are used, this can be a computationally expensive task involving an exhaustive search across all prospective polygons. In this paper, we propose a methodology that exploits kd-trees as an efficient nearest neighbour search algorithm to significantly reduce the effective number of polygons being searched and expedite the lookup process. The kd-tree is built from either the polygon centroids and/or carefully chosen other points within the polygons. We further demonstrate a successful hybrid strategy by combining a range search with the tree based ranking. Our code has been made publicly available as the R package *RapidPolygonLookup*.

Keywords: Polygon lookup · r · Spatial · kd-tree · Visualization

1 Introduction

Finding underlying patterns in spatial data is a challenging task and often requires visual inspection of the data points. Typically, such spatial visualizations are created on some sort of a reference background, e.g., if the data are demographic then they are overlayed on maps of corresponding geographical entities like cities, states, or regions. Contextualizing the points in such a way helps derive insights regarding the data generation process and better understand the underlying spatial structure.

Consider, as an example, the data shown in Fig. 1. The figure on the left shows the zinc concentration levels in a flood plain of the Meuse river close to the village of Stein [14, 16]. While such a graph with minimal clutter showcases the locations of interest neatly, it lacks a spatial context that would otherwise be useful for deriving actionable insights. A lot of modern data sets are now geo-tagged and contain location information at a latitude-longitude level. Cell phone usages, vehicle tracks, demographics, crime, and flu outbreaks are some

© Springer International Publishing Switzerland 2015
A. Plakhov et al. (Eds.): EmC-ONS 2014, CCIS 499, pp. 192–206, 2015.
DOI: 10.1007/978-3-319-20352-2_13

examples of data that have a spatial attribute attached to them. While in certain cases the scenario on the left panel of Fig. 1 may be preferred, the rather limiting nature of such an exploration can be overcome by graphing the same points on the associated map background as shown in the right panel of the same figure.

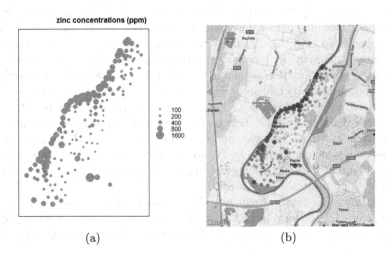

(a)   (b)

Fig. 1. (a) zinc concentration levels in the Meuse river from [14,16], (b) contextualizes the same information with a background of the surrounding areas as generated by the package *RgoogleMaps* [7]

Though such preliminary visualizations of spatially dispersed points are a great starting point, they need to be supplemented with additional graphical infrastructure for more robust analyses. For example, in many cases we are faced with the task of evaluating spatial patterns with respect to a reference point, e.g., variation of crime patterns across the blocks of a city, traffic density around major intersections, real estate prices across school districts etc. Analyzing spatial data with such context proves to be a useful tool for making informed decisions, designing public policies, and efficiently allocating resources.

We claim that even the formulation and generation of statistical hypothesis to test can be greatly facilitated by the additional layer of information provided by a map.

1.1   Spatial Aggregation

A concise overview of the relevance of the spatial aggregation problem in general, and specially in economics, is given in [5]. In empirical spatial data analysis, this problem is often referred to as the *modifiable areal unit problem* (MAUP)[10,12]. In this paper, we ignore the statistical consequences of coarsening the resolution of spatial data (see, e..g [11]) and instead focus on the algorithms that can be used to **efficiently** map spatially dispersed data points to higher geographical entities like zip codes, Census blocks, or districts. Once associated with a particular spatial region, information from all points within a region can be aggregated to search for patterns and derive meaningful insights.

The particular type of aggregation we have in mind pertains to (stratified) counting of spatially referenced points that fall within arbitrarily shaped polygons that form a complete spatial partition, i.e. leave no holes nor overlap in any way. Without loss of generality we assume that every search point lies within a unique polygon. In case these assumptions are not met, it should be straightforward to either augment the search area with polygons or eliminate intersections.

The case of rectangular regularly shaped polygons is easily achieved and poses no computational challenges. However, mapping arbitrary points to irregularly shaped polygons can be a computationally expensive exercise that typically involves looking up each data point in multiple spatial polygons until the search criteria are met. This can become a daunting task if the number points to be mapped, which we denote by N, go beyond a few thousand. In particular, the overall computational cost is expected to be of $O(N \cdot M)$ for M polygons, with a very high multiplicative constant.

Throughout this text we will use US Census block polygons [1] as a well-suited example for the spatial partitions that our proposed algorithm is tailored to. Figure 2b illustrates an overlay of these census polygons for the city of San Francisco.

(a)   (b)

Fig. 2. (a) Crime incidents in San Francisco; the red points indicate violent crimes and the green ones denote non-violent crimes, (b) Overlay of the US 2010 Census blocks on crime incidents in San Francisco (Color figure online).

The Figure also shows 10,000 crime incidents that occurred in San Francisco in 2012 [17]. The red points on the figure denote violent crimes, which we have defined as assault, robbery, rape, kidnapping, and purse snatching. The green points are non-violent crimes. The left plot in this figure shows the spatial distribution of crime across the entire city. The plot on the right (b) adds more context to the crime incidents by visually mapping each incident to a Census block. Both panels were generated by the package *RgoogleMaps* [7]. This association can aid efficient allocation of policing resources and provide insights required for formulating demographic policies.

In this paper, we propose a methodology to expedite the search process using efficient data structures by exploiting k-d trees as a fast nearest neighbours algorithm to dramatically reduce the number of prospective polygons in the search space. Leveraging k-d trees, we compute a list of polygons that are most likely to contain each data point. The provision of such a list, ordered by likelihood, reduces the number of effective polygons that need to be exhaustively searched to determine the inclusion of the point in consideration.

Note that we neither compare our method with *R-trees* [6] nor do we claim algorithmic superiority over those data-structures that are specifically designed to deal with objects of spatial extent. Our reasoning is more pragmatic and motivated by code maturity and availability: well-tested, well-documented, easily usable and fast open source k-d tree libraries exist in several programming languages. R-trees on the other hand are more recent and have been embraced by a smaller community which has resulted in much sparser code availability. In addition, searches on trees that are based on hierarchical partitioning of rectangles would still necessitate the point-in-polygon operation on the remaining candidates. We speculate that R-trees could potentially speed up our range search which we describe in Sect. 3.2.

1.2   Point in Polygon

Determining the inclusion of a point q in a 2D planar polygon P is a geometric problem that has been studied in detail. Two commonly used methods are [13,18]:

- The *Winding Number* (Z_{wn}) method, which counts the number of times the polygon winds around the point q. The point is outside only when this "winding number" $Z_{wn} = 0$; otherwise, the total turn must be a whole number of revolutions, $Z_{wn} = 2n\pi$, and the point is inferred to be inside.
- The *Crossing Number* (Z_{cn}) method, which counts the number of times a ray starting from the point q crosses the polygon boundary edges. The point is outside when this "crossing number" is even; otherwise, when it is odd, the point is inside. This method is sometimes referred to as the "even-odd"test or the "ray casting" method.

It should be easy to see that both algorithms scale linearly in the number of vertices n of the respective polygon, i.e. O(n). The claim made in [13] that the ray crossing algorithm is more than twenty times faster than the winding number method is refuted in [18] (Fig. 3).

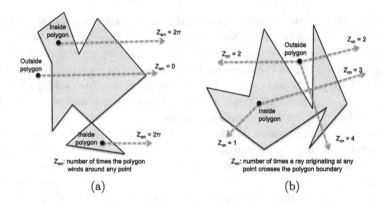

Fig. 3. Illustration of (a) the winding number concept: at a given point q "watch" another point p completely traverse the polygon boundary ∂P and keep facing point p. If $q \in P$, the total angular turn would be a multiple of 2π otherwise it would be exactly zero; and (b) the crossing number idea: we draw a "ray" R from the point in question q in an arbitrary direction and count the number of intersections of R with ∂P. The point q is in or out of P if the number of crossings is odd or even, respectively [13].

1.3   Polygon Ranking

Execution of point-in-polygon can be expensive if the number of prospective polygons is large. One way economize on this is to reduce the number of prospective polygons by ordering them based on the likelihood of inclusion of the point. The motivation behind our paper lies around exploring methods for rank ordering polygons. Here we provide a brief description of some "simpler" rank ordering methods that will help lay the stage for our methodology explained later.

Exhaustive Search: Randomly pick the polygons and check until a match is found. This process is "exhaustive" in the sense that the search is executed across the entire polygon space until the success criteria are met.

Area Based Ranking: Order the polygons based on the polygon area. This could prove to be a fast and effective method if the points are evenly spread out and the polygons are regularly shaped.

Prior Assignments: Rank the polygons based on the number of points that have already been mapped to them. If the data generation process of the points is consistent, then the previous mappings can be "cached" for the remaining points.

Least Number of Vertices: The cost of executing point-in-polygon increases with the number of vertices of the polygon. Rank ordering the polygons starting with those with the least number of vertices can then help reduce the overall execution time.

While the methods mentioned above do provide a starting point, they fail to scale and generalize if the number of data points to be mapped is large and if the polygons are irregularly shaped. Our paper tries to address these problems. This paper is organized as follows: in the next section we lay the background of our methodology by providing a brief description of k-d trees and related concepts. Section 3 details our approach and Sect. 3.4 provides the testing benchmarks. Section 4 discusses a few potential examples where our algorithm can be applied and Sect. 5 concludes the paper.

2 K-D Trees

K-D trees are data structures that partition a k-dimensional space to store a set of finite points [3]. K-d trees are binary trees such that each node has a maximum of two child nodes. The entire tree has a guaranteed depth of $log_2(N)$ where N is the number of points in the set. They provide an efficient mechanism to search a point in a k-dimensional space by hierarchically decomposing the space into smaller and smaller partitions. This hierarchical decomposition is achieved by recursively partitioning the space using only one dimension at a time and then recycling once the kth dimension is reached. A point to be looked up can then be dropped down the tree starting at the root node and the condition for inclusion can then be checked at each node that falls on the designated path.

A popular and useful application of k-d trees is searching for the nearest neighbours by leveraging the structure of a tree. This is possible because the structure allows for quick checking and elimination of large partitions in the search space.

For example, a quick first approximation can be found by traversing the point down the tree until a leaf node is reached. This first approximation can be described as the "current best". Consequently, if there is any point that is closer to the search point, it has to lie within the hypersphere drawn with the search point at the center and the distance between the search point and the current best as the radius. Given that the hyperplanes partitioning the space are axis aligned, it can easily be checked if the hypersphere and the hyperplane intersect. If not, the partition on the other side of the hyperplane can be entirely eliminated from the search space. On average, such a procedure is at least $O(log(N))$, where N is the number of nodes in the tree, since at least one leaf of the tree needs to be visited. At max, this value can be N such that each node is visited at most once [9].

3 Methodology

Our methodology relies on the efficient data structures that k-d trees produce. We use these structures to compute nearest polygon neighbours for any given search point and then check which of these polygon neighbours contains the search point. We economize on the computation time by significantly reducing the number of effective polygons that need to be searched to determine the parent polygon of the data point.

3.1   Polygon Centroid Method

Consider a search point s in a k-dimensional space that can be represented by $(s_1, s_2, \ldots, s_k)$. Additionally, consider P polygons in a similar k-dimensional space. We determine the centroid for each of these P polygons such that we effectively have P points in the k-dimensional space which can be represented as $(p_1, p_2, \ldots, p_k)$.

Using the centroids of the polygons we construct a k-d tree and then use the tree to compute M (typically M is chosen to be around 10) nearest neigbours for the search point [2]. This computation provides an ordered list of M polygons which are most likely to include the search point. We run an exhaustive check on this ordered list to find the polygon which contains the point in consideration. Exploiting k-d trees in such a manner allows us to significantly reduce the number of effective polygons that need to be exhaustively searched to determine which polygon does the point belong to. Figure 4 suggests that indeed for the vast majority of search we needed to test only $1 - 2$ polygons.

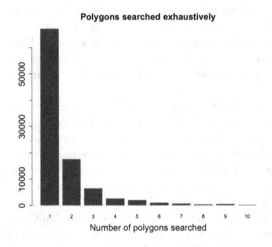

Fig. 4. Distribution of the number of polygons tested to check for inclusion of search points.

However, we also notice a (small) percentage of cases where the nearest $4-10$ centroids turned out to not include the correct polygon. Such exceptions warrant closer inspection and Fig. 5a shows a few of these cases. It can be seen that many points lie on the edges of the candidate polygons thereby making it less likely for the closest centroid to belong to the parent polygon. These exceptions, however, can be minimized by increasing the number of nearest neighbours being computed. Increasing the number of neighbours provides a longer list of candidate polygons for exhaustive search, which typically leads to more points being mapped to polygons at the cost of a modest increase in computation time. Figure 6 shows the proportion of points left unmapped in this process. We see that as the number of contender polygons is increased by computing more nearest neighbours, the proportion of search points left unmapped gradually declines. The current code [8] employs a hybrid strategy by executing a range search (described in the next section) on the remaining unmapped points such that no point remains unassigned.

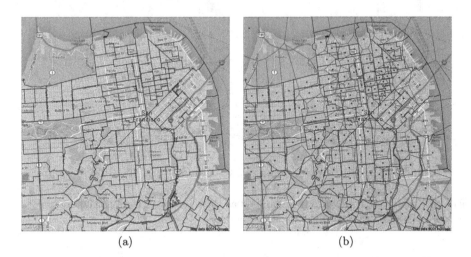

(a)   (b)

Fig. 5. (a) Crime incidents that could not be mapped to the ten "closest" polygons - as measured by the distance to the centroids. The lower and upper cluster of points appear inside larger polygons that are surrounded by several smaller ones. (b) Polygons with their centroids (green) as well as an overlay of the Voronoi cells (red) of the polygon centroids (Color figure online).

3.2   Range Search

For each polygon we initially compute the *minimum bounding rectangle* (MBR). The main search idea then is simply to rule out those polygons whose MBR does not contain the search point. The ranking within the remaining polygon candidates can be chosen among the following heuristics: (i) closest centroid, (ii) closest vertex or (iii) largest area. We note in passing that R-trees [6] also utilize the idea of bounding box intersection with search objects.

Whether the range search is slower or faster than the closest-centroid ordering depends very much on the number of polygons searched before inclusion is decided. Currently, the user chooses the maximum number of nearest centroids k (returned by the k-d tree) to search before reverting to the range search. Each data set will likely possess a unique optimal value for k which can be estimated by experimenting with a small sample; we therefore do not offer strong recommendations for choices for k. The next section introduces an alternative search strategy which suggests a maximum value of $k = 4$.

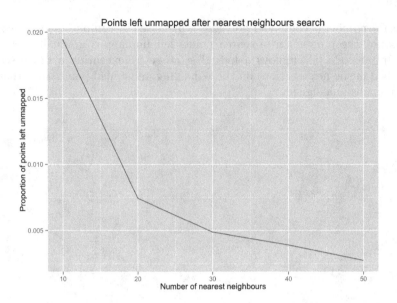

Fig. 6. Proportion of points left unmapped as the number of nearest neighbours computed is changed for a given set of search points

3.3 Closest Vertex

We believe that representing polygons by their centroids and searching for the nearest centroid(s) for each earch point s can be improved upon. To motivate the next idea, we inspect those points for which the 10 nearest polygon centroids did not lead to a succcessful match as graphed in Fig. 5a.

It is evident that points closer to the edges of a polygon are less likely to be associated with the correct centroid than interior points. Constellations of large polygons surrounded by many small ones lead to regions where the "correct" centroid is not even in the list of the ten closest; the two clusters of unmapped points in Fig. 5a exemplify that situation.

To better understand which search points are associated with which polygons, we compute and overlay the Voronoi tesselation for the set of polygon centroids as shown in Fig. 5b. Ideally, the Voronoi cells would coincide closely with the actual polygons, in which case the closest centroid would be the correct match. However, we notice frequent substantial mismatch between these geometric entities. Let us define the "Voronoi coverage"of a polygon as the fraction of its area covered by its own Voronoi cell. Figure 7a shows its distribution which is very heterogenous.

Summing up the weighted proportions yields a total area covered of 60.5 %. Given a random point from a spatially uniform distribution (rarely true for real data sets, of course), the closest centroid is correct if and only if that point happens to be inside the "Voronoi coverage". Hence, the probability that the nearest centroid is a successful match should be about 60.5 %.

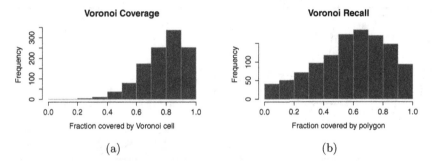

Fig. 7. (a) For each polygon we define the "Voronoi coverage" as the fraction of its area covered by its own Voronoi cell. For values close to 1 the nearest centroid to a search point would (almost) always indicate its enclosing polygon. Note the large fraction of polygons with coverage less than 80 % which leads to the observed inefficiencies in the centroid based algorithm. (b) "Voronoi recall rate"(true positives) defined as the fraction of its area covered by its respective polygon.

A related, and perhaps more relevant, measure is the conditional probability of point q belonging to polygon p_i given that q is inside its Voroni cell V_i: $P(q \in p_i | q \in V_i)$. In a Bayesian setting this would be the posterior probability resulting from updating existing prior probabilities $P(q \in p_i)$. We compute this "Voronoi recall rate"as the fraction of its area covered by its respective polygon and graph its distribution in Fig. 7b. Again, we see a greatly varying distribution.

One idea to improve these odds is to augment the set of centroids with additional representative points, either randomly chosen or in an adaptive fashion where the falsely classified regions would be preferentially selected. Note, however, that this is a difficult global optimization problem in which local improvements of coverage in one polygon can have a detrimental effect on its neighbours.

Instead, we propose to represent the polygons not by their centroids but their vertices. It should be intuitively clear that a search point is closest to those vertices "surrounding it", i.e., those that belong to the enclosing polygon. Note that we are now relying strongly on our initial assumption that the given set of polygons constitutes a complete partition of the area of interest; the polygons neither overlap nor leave holes. In that case most vertices are shared by at least two polygons. In the case of corners, they can be shared by three or more. Experiments with the Census block polygons from the US Census 2010 [1] confirm this expected behavior resulting in an average number of about 1.6 polygons tested per point. While this is already a notable improvement over the closest-centroids method, we can further improve the ranking within the list of equidistant vertices and their corresponding polygons by the following heuristic. We hypothesize that for most combinations of polygons and search points the correct centroid lies "on the other side"of the point than the closest vertex whereas the centroid of the incorrect, neighbouring polygon tends to be on "same side". Figure 8 visualizes this idea via two examples from the census blocks.

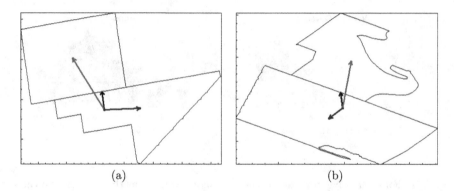

(a)   (b)

Fig. 8. Two examples for the direction of the vectors originating from a randomly chosen search point s (black dot, origin of all three arrows) to (i) the closest vertex (red point, black -shortest- arrow), (ii) the centroid of the enclosing polygon (blue arrow, second shortest) and (iii) the centroid of the neighboring polygon (green arrow -longest). Both examples support the proposed heuristic that the angle between black-blue vectors tend to be larger than between black-green vectors (Color figure online).

In both cases the closest vertex (red point) to the search point (black point) belongs to the two polygons drawn. We would like to rank the matching polygons (blue centroids) ahead of the "wrong" polygons (green centroids). The notion of "other/same side" is measured by the angles between the black vector and the blue/green vectors respectively. Our heuristic is simply to break the rank tie between the two polygons by increasing the prior probability of the one with the larger angle.

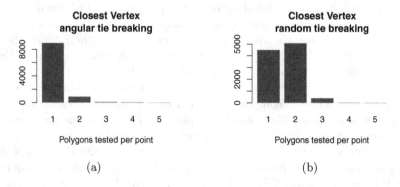

(a)   (b)

Fig. 9. Distribution of the number of times the *point.in.polygon()* function must be executed before a match is found for 10, 000 randomly selected crime locations in San Francisco. Ties in vertex-point distance are broken: (a) via the described angular heuristic and (b) randomly.

Figure 9a shows the successful application of this simple idea: the polygons ranked number one are correct in about 90 % of the cases for a test set of 10^4 points.

Breaking ties among the polygons randomly instead of the "angular heuristic" leads to an approximately 50 % chance of choosing correctly between the two candidates, as shown in Fig. 9b.

We speculate that for arbitrary sets of polygons the simple strategy of closest vertex/centroid must likely be modified to retain its efficiency. The current termination of the search upon the first successful match needs to be refined in case of overlapping polygons which would allow for multiple assignments to the respective enclosing geometries. Limiting the polygon candidates with the above mentioned range search (that we do not explain in detail here) and ranking the polygons within that subset would likely be a near optimal hybrid algorithm. A violation of our second assumption – the existence of "holes", i.e. regions in space that are not covered by any polygon – can be dealt with by this same approach or by a pre-processing step which would create additional space-filling polygons.

We use crime incidents from San Francisco [17] to benchmark our algorithm. For polygons, we use Census blocks from the US Census 2010 [1]. The results of the benchmarks are provided in Fig. 10. For all three proposed methods Fig. 10 shows a near linear scaling in computation time as the number of search points are increased. These results were produced after averaging the run times over 5 iterations. Additionally, for each point in the data set, we record the number of polygons that were exhaustively searched before the parent polygon was found. This was done for each of the three methods.

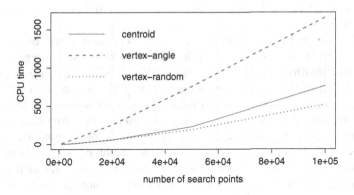

Fig. 10. CPU time taken to map SF crime incidents to Census block polygons for the three algorithms proposed in the paper: (i) closest-centroid method as well as closest vertex with (ii) angle heuristic and (iii) random ranking.

3.4 Benchmarks

On average, we had to execute the point-in-polygon check:

1. 1.7 times for the closest-centroid method,
2. 1.6 times for the closest-vertex, random tie breaking method,
3. 1.12 times for the closest-vertex, angular heuristic method.

These results show a considerable reduction in the number of effective polygons searched, highlighting the efficiency gains proposed by our methodology. Surprisingly, the method with the lowest average execution of the polygon inclusion test emerges as the slowest so far as Fig. 10 suggests. We speculate that the time taken be the computation of the necessary angles exceeds the small overhead of more polygon tests.

In forthcoming work, we will provide benchmarks for a larger variety of data sets in combination with different types of polygonal partitions.

All analyses were performed using the open source R statistical programming language [15].

4   Further Examples

The aggregation of spatial points to higher geographic entities is a generic computational task that finds applicability in various domains. Consider, as an example, the case of conflict outbreaks, events that have both spatial and temporal characteristics much like point processes. For example, Weidmann and Ward [19] try to predict conflict in Bosnia at a municipality level using a spatio-temporal regression model. It is easy to see how this process can be easily generalized to a different hierarchical structure as required by the governing bodies using the spatial aggregation methodology we present in this paper. Historical geocoded data of conflict outbreaks can be mapped to any spatial polygon structure that can further be joined with local demographic, social, and economic variables to gain deeper insights.

A more practical application in an era of social media and real-time data collection could include estimation of demand for taxi services at a grid or block level. Such estimation would involve aggregating historical pickup locations to most relevant spatial partitions, combining them with more contextual information like proximity to commercial establishments, residence spaces, larger travel stations, and generating in-situ predictions for different time divisions of the day.

Another example where aggregating spatial points could prove helpful is predicting forest fires. Forest fire locations can be combined with information on temperature, wind, humidity, and vegetation along with variables such as the fire weather index [4] to find fire hot-spots (areas that are more prone to destructive fires), and consequently facilitate proactive intervention where relevant.

5   Conclusion

In this paper, we explore an efficient spatial aggregation technique that can be used to assign spatially dispersed points to common geographical entities. We exploit k-d tree data structures to compute approximate nearest neighbours for search points by computing distances between the coordinates of the points and a set of fixed points which represent the polygons. Choices for the latter investigated in this paper are (i) the polygon centroids and (ii) the vertices. Such a procedure provides efficiency gains as the results from nearest neighbours search significantly reduce the number of *point.in.polygon()* function calls. Combining these tree based methods with a range search as pre- or post- processor makes our method widely applicable even when the polygons do not provide a complete spatial partition. Our code has been made publicly available as an R package [8].

In future work we plan to augment the polygons with auxiliary information such as convexity, largest enclosing circle, and/or rectangle, all of which can lead to even more efficient point to polygon mappings.

References

1. Almquist, Z.W.: Us census spatial and demographic data in R: TheUScensus2000 suite of packages. J. Statis. Softw. 37(6), 1–31 (2010). http://www.jstatsoft.org/v37/i06/
2. Arya, S., Mount, D., Kemp, S.E., Jefferis, G.: RANN: Fast Nearest Neighbour Search (wraps Arya and Mount's ANN library) (2013). http://CRAN.R-project.org/package=RANN. (r package version 2.3.0)
3. Bentley, J.L.: Multidimensional binary search trees used for associative searching. Commun. ACM 18(9), 509–517 (1975). http://doi.acm.org/10.1145/361002.361007
4. Cortez, P., Morais, A.d.J.R.: A data mining approach to predict forest fires using meteorological data. In: 13th Portuguese Conference on Artificial Intelligence, New Trends in Artificial Intelligence, pp. 512–523. Associao Portuguesa para a Inteligncia (2007)
5. Dusek, T.: Spatially aggregated data and variables in empirical analysis and model building for economics. Cybergeo: Eur. J. Geogr. (2004). http://cybergeo.revues.org/2654 (dossiers, 13ème Colloque Européen de Géographie Théorique et Quantitative, Lucca, Italie, 8–11 septembre 2003, document 285)
6. Guttman, A.: R-trees: A dynamic index structure for spatial searching. In: Proceedings of the 1984 ACM SIGMOD International Conference on Management of Data, SIGMOD 1984, pp. 47–57. ACM, New York (1984). http://doi.acm.org/10.1145/602259.602266
7. Loecher, M.: RgoogleMaps: Overlays on Google map tiles in R (2013). http://CRAN.R-project.org/package=RgoogleMaps. (r package version 1.2.0.5)
8. Loecher, M., Kumar, M.: RapidPolygonLookup: Polygon lookup using kd trees (2014). http://CRAN.R-project.org/package=RapidPolygonLookup. (r package version 0.1)
9. Moore, A.: Efficient memory-based learning for robot control. Ph.D. thesis (1991)
10. Openshaw, S.: The modifiable areal unit problem, vol. 38. Geo Books, Norwich (1983)
11. Openshaw, S.: Ecological fallacies and the analysis of areal census data. Environ. Planning A 16(1), 17–31 (1984)
12. Openshaw, S., Taylor, P.J.: A million or so correlation coefficients: three experiments on the modifiable areal unit problem. Stat. Appl. Spat. Sci. 21, 127–144 (1979)
13. O'Rourke, J.: Computational geometry in C. Cambridge University Press, Cambridge (1998)
14. Pebesma, E.J., Bivand, R.S.: Classes and methods for spatial data in R. R News 5(2), 9–13 (2005). http://CRAN.R-project.org/doc/Rnews/

15. R Core Team: R: A Language and Environment for Statistical Computing. R Foundation for Statistical Computing, Vienna, Austria (2013). http://www.R-project. org/
16. Bivand, R.S., Edzer Pebesma, V.G.R.: Applied spatial data analysis with R. Springer, New York (2013). http://www.asdar-book.org/
17. San-Fransico-Government: San fransico police department (sfpd) crime incident data (2013). https://data.sfgov.org/, calendar-year data can be extracted from https://data.sfgov.org/Public-Safety/SFPD-Reported-Incidents-2003-to-Present/ dyj4-n68b
18. Sunday, D.: Inclusion of a point in a polygon (2014). http://tinyurl.com/q4f6dgs
19. Weidmann, N.B., Ward, M.D.: Predicting conflict in space and time. J. Conflict Resolut. **54**(6), 883–901 (2010)

Author Index

Printed in the United States
By Bookmasters